# COMMAND CONCEPTS

## A THEORY DERIVED FROM THE PRACTICE OF COMMAND AND CONTROL

*Carl H. Builder*
*Steven C. Bankes*
*Richard Nordin*

Prepared for the
Office of the Secretary of Defense

**National Defense Research Institute**

**RAND**

The research described in this report was sponsored by the Office of the Secretary of Defense (OSD), under RAND's National Defense Research Institute, a federally funded research and development center supported by the OSD, the Joint Staff, the unified commands, and the defense agencies, Contract DASW01-95-C-0059.

**Library of Congress Cataloging-in-Publication Data**

Builder, Carl H. , 1931- 1998.
   Command concepts : a theory derived from the practice of
command and control / Carl H. Builder, Steven C. Bankes, Richard
Nordin.
      p.      cm.
   "Prepared for the Office of the Secretary of Defense by RAND's
National Defense Research Institute."
   "MR-775-OSD."
   Includes bibliographical references (p. ).
   ISBN 0-8330-2450-7
   1. Command and control systems.   2. Command of troops—Case
studies.   I. Builder, Carl H.   II. Bankes, Steven C.   III. Richard
Nordin.   IV.  United States.  Dept. of Defense.  Office of the
Secretary of Defense.   V.  National Defense Research Institute
(U. S. ).   VI.  Title.
UB212.N67     1999
355.3 ' 3041—dc20                                              96-39019
                                                                          CIP

RAND is a nonprofit institution that helps improve policy and decisionmaking through research and analysis. RAND® is a registered trademark. RAND's publications do not necessarily reflect the opinions or policies of its research sponsors.

Published 1999 by RAND
1700 Main Street, P.O. Box 2138, Santa Monica, CA 90407-2138
1333 H St., N.W., Washington, D.C. 20005-4707
RAND URL: http://www.rand.org/
To order RAND documents or to obtain additional information,
contact Distribution Services: Telephone: (310) 451-7002;
Fax: (310) 451-6915; Internet: order@rand.org

The qualities of commanders and their ideas are more important to a general theory of command and control than are the technical and architectural qualities of their computers and communications systems. This theory separates the art of command and control (C2) from the hardware and software systems that support C2. It centers on the idea of a *command concept,* a commander's vision of a military operation that informs the making of command decisions during that operation. The theory suggests that the essential communications up and down the chain of command can (and should) be limited to disseminating, verifying, or modifying command concepts. The theory also suggests, as an extreme case, that an ideal command concept is one that is so prescient, sound, and fully conveyed to subordinates that it would allow the commander to leave the battlefield before the battle commences, with no adverse effect upon the outcome.

This report advances a theory about military command and control. Then, through six historical case studies of modern battles, it explores the implications of the theory both for the professional development of commanders and for the design and evaluation of command and control architectures. The report should be of interest to members of the Joint Staff and of the services involved in developing command and control doctrine for the U.S. military, and to all those interested in the "military art and science" of command and control.

This research was performed under the project "Warfare in the Information Age," within the Acquisition and Technology Policy Center of RAND's National Defense Research Institute, a federally funded research and development center sponsored by the Office of the Secretary of Defense, the Joint Staff, the unified commands, and the defense agencies.

# CONTENTS

# FIGURES

## A DEEPER THEORY OF COMMAND AND CONTROL IS NEEDED

In an age of abundant, almost limitless, information and communications capabilities, decisionmakers are increasingly faced with the problem of too much information, rather than too little. In today's information-oriented society, winnowing, filtering, correlating, and fusing information have become as important as acquiring the information, or (regrettably) even as important as its content, if not more so. Understanding what information is most essential for decision-making—so that the information being communicated, processed, or displayed can be bounded—is now a major issue in the design of computer-aided decision support systems.

Nowhere has the problem of overabundant information become more apparent than in military command and control, where the accelerating technologies of communications and computers[1] have flooded commanders at all levels with so much information that they sometimes seem no longer able to digest or comprehend it. The prevailing approach to this problem is to apply still more technology in

---

[1]Nothing on the face of the earth is changing more rapidly than the numbers and capacities of electronic communications networks. The capacity of the communications networks available to the public has been doubling every two or three years for more than a decade and is likely to continue that pace for more than a decade to come. For some of the broader implications of this extraordinary change, see Carl H. Builder, "Is It a Transition or a Revolution?" *FUTURES: The Journal of Forecasting, Planning and Policy,* Vol. 25, No. 2, March 1993, pp. 155–168.

the form of computers and software, in order to sort through, filter, and display the information in ways that will assist the commander in focusing on the "right" information. Of course, this approach assumes that the commander and his responsibilities, circumstances, and decisions are understood well enough for his information needs to be anticipated.[2] It also assumes that the C2 issues for commanders are acquiring enough information, sorting through it, and then maintaining connectivity with subordinates so that they can be directed.

How technology could or should be applied to the issue of command and control has been addressed by a number of writers, but contemporary *theories* about command and control (C2) are, by and large, theories about organizations and communications. Such theories often take the form of charts or diagrams: As organizational charts, they relate commanders to the things they control; as network-wiring diagrams, they relate communications nodes and links in terms of their informational functions or capacities. In these theories, commanders are the occupiers of boxes, and what flows through the channels is messages.

But such theories do not explain what C2 does and does not do, can and cannot do. Perhaps most importantly, they do not explain the qualities of the ideas or how those ideas are expressed within the system. A comprehensive theory of C2 should explain not only how to organize, connect, and process information, it should also explain something about the quality of ideas and their expression and about how the qualities of people contribute to or detract from C2, not just how they should be organized and wired together. What is needed is a deeper theory that encompasses the high-level, creative aspects of command as well as the direct-order and control aspects.

A theory of C2 should explain how C2 systems, including commanders, should work and the ideal circumstances in which that work can occur, and should provide performance measures for commanders and their staffs, as well as for the communications and computers that support them.

---

[2]In this report, we use *he/his* throughout for clarity, not to imply gender significance.

Kenneth Allard, in commenting on the following definitions of *command, command and control,* and *command and control system* in the *Department of Defense Dictionary of Military and Associated Terms,* notes that "one of the most striking characteristics of these definitions is the extent to which they evoke the personal nature of command itself, especially the fact that it is vested in an individual who, being responsible for the 'direction, coordination, and control of military forces,' is then legally and professionally accountable for everything those forces do or fail to do":[3]

> *Command:* "The authority vested in an individual of the armed forces for the direction, coordination, and control of military forces."
>
> *Command and control:* "The exercise of authority and direction by a properly designated commander over assigned forces in the accomplishment of the mission. Command and control functions are performed through an arrangement of personnel, equipment, communications, facilities, and procedures which are employed by a commander in planning, directing, coordinating, and controlling forces and operations in the accomplishment of the mission."
>
> *Command and control system:* "The facilities, equipment, communications, procedures, and personnel essential to the commander for planning, directing, and controlling operations of assigned forces pursuant to the missions assigned."[4]

When we use these three terms, the above definitions are implied in them.

Going beyond personality alone, our theory suggests that the essence of command lies in the cognitive processes of the commander—not so much the way certain people do think or should think as the ideas that motivate command decisions and serve as the basis for control actions: Ideally, the commander has a prior concept of impending

---

[3]Kenneth Allard, *Command, Control, and the Common Defense,* Washington, D.C: National Defense University, Institute for National Strategic Studies, revised 1996, pp. 16–17.

[4]U.S. Joint Chiefs of Staff, *Department of Defense Dictionary of Military and Associated Terms,* Washington, D.C.: Office of the Joint Chiefs of Staff, JCS Pub. 1, January 1986, p. 74 (quoted in Allard, 1996 rev., p. 16).

operations that cues him (and his C2 system) to look for certain pieces of information.

Our theory cuts through the technological overlay[5] that now burdens the subject, and can be used as a template to reexamine some familiar historical instances of C2 success and failure. The theory represents an attempt to separate the intellectual performance of the commander from the technical performance of the C2 system.

## COMMAND CONCEPTS

Looking across the history of military operations and considering the substance of communications between commanders and their subordinates, we noted that one of the most consistently evident topics is some vision of the conduct of a military operation: what could and ought to be done in applying military force against an enemy. Renowned commanders are concerned mostly with explaining and asking after their vision or expectations of possible and desirable operations: "Are things going as we planned (envisioned)? If not, what is broken and needs fixing? Why and where are things going wrong? Is the plan (vision) wrong, or does it simply need some adjustment?"

The source of such visions, of course, lies inside human minds—in complex sets of ideas that might be called "command concepts." Evidence of command concepts is found mostly in war and battle plans, sometimes in the setting of military objectives, less often in the deployment and commitment of forces, and perhaps least often in the issuance of direct orders.

We define a *command concept* as a vision of a prospective military operation that informs the making of command decisions during that operation. If command concepts are the underlying basis for command, then they provide an important clue to the minimum essential information that should flow within command and control systems. If what flows through command and control networks is (or ideally should be) command concepts, then informational needs can be bounded. Rather than creating a C2 system that can transmit all

---

[5]The technological overlay is mostly from communications and computers, which are changing at a remarkable pace. Command and control itself is conducted mostly in and through human minds, which change much more slowly.

the information that can be acquired, or all that the bandwidth will bear, an ideal C2 system would transmit *only information that helps the commander convey his command concept, or alter it.* From that perspective, reporting the number of vehicles in a battle, for example, is pertinent only if it is somehow relevant to the command concept or how the concept will evolve. And since commanders do not need to adopt a new concept until it is clear to them that the existing one has failed or can be bettered, what a commander must hear, at the minimum, is information that disturbs or refutes his concept, even though he probably wishes to hear its confirmation.

The following is a list of elements that should be found in an ideal command concept:

- Time scales that reveal adequate preparation and readiness, not just of the concept but of the armed forces tasked with carrying out that concept.

- Awareness of the key physical, geographical, and meteorological features of the battle space—situational awareness—that will enable the concept to be realized.

- A structuring of forces consistent with the battle tasks to be accomplished.

- Congruence of the concept with the means for conducting the battle.

- What is to be accomplished, from the highest to the lowest levels of command.

- Intelligence on what the enemy is expected to do, including the confirming and refuting signs to be looked for throughout the coming engagement.

- What the enemy is trying to accomplish, not just what his capabilities and dispositions may be.

- What the concept-originating commander and his forces should be able to do and how to do it, with all of the problems and opportunities—not just the required deployments, logistics, and schedules, but the nature of the clashes and what to expect in the confusion of battle.

- Indicators of the failure of, or flaws in, the command concept and ways of identifying and communicating information that would change or cancel the concept.

- A contingency plan in the event of failure of the concept and the resulting operation.

Finally, if the notion of command concepts has validity, it should apply to all levels of command—from theater commanders to squad leaders, each of whom will have his own command concept that forms the basis for his decisions. Each of those concepts should be hierarchically nested and consistent. That is, the squad leader understands his platoon leader's concept for platoon operations and has then developed his own for squad operations, which is supportive of the concept for the next higher level and consistent with it.

## CASE HISTORIES

In theory, therefore, if a commander's vision of battle was sound and was fully conveyed to subordinates beforehand, there would be no need for information to be in the C2 system during the ensuing battle. Conversely, the theory suggests that needing a given amount of information in the C2 system during the battle relates directly to failures associated with the validity or completeness of the command concept or its clear conveyance to subordinates.

We look at this theory through the lens of six battles drawn from modern military history. In each of these historical cases, we look for the existence, clarity, and expression of a command concept. We ask if the command concept was valid and adequately supported by the C2 system. If no explicit concept, such as an operational plan, exists, we infer an ideal concept from information on all aspects of the battle.

We also chose our examples to illustrate instances in which:

- The command concept was either ideal or unsound for the circumstances presented to the commander.

- The expression of the concept by the commanders was either complete—that is, the expressed concept contains all the elements listed in the preceding section—or incomplete.

- The C2 system was either adequate or inadequate to support the commander and his command concept.

Collectively, these six cases support and illuminate our theory: Admiral Chester Nimitz at Midway and General Douglas MacArthur at Inchon provide near-ideal examples of a valid command concept, completely expressed, and adequately supported by their C2 systems.  Field Marshal Bernard Montgomery at MARKET-GARDEN provides an example of a command concept that, although clearly expressed, was structurally unsound because it tried to fit the operation to the available air forces rather than deciding which service or force would be best after carefully researching the plan.  In addition, it was not adequately supported by a C2 system that could correct its errors, and it ignored correct intelligence.  Lieutenant Colonel Harold Moore at Ia Drang and General Heinz Guderian at Sedan, for different reasons, did not have clearly expressed command concepts. They relied on their doctrine and training instead; and their C2 systems, while otherwise adequate, could not substitute for the strategic decisions that had to be made and communicated.  General H. Norman Schwarzkopf in DESERT STORM had an adequately detailed and expressed command concept; when events required him to accelerate his plans, his C2 system served him well in alerting him to those events.

## CONCLUSIONS

If it can be demonstrated that command concepts are (or ideally should be) the essential substantive content of the information flowing through command and control systems, then a powerful theory of command and control is indicated.  Ideal commanders and ideal command concepts are, of course, only a reference point, rarely if ever observed in war.  But they can serve as a useful reference point for a theory of command and control that:

- Defines the highest-priority information—what command and control systems must be designed to handle most quickly and with the highest fidelity.

- Separates the proper *intellectual* burdens of the commander from the *communication* burdens of the command and control system, both before and during the battle.

- Makes the commander, not his C2 system, responsible for the quality of his ideas, his ability to express them, and his receptivity to information that disturbs or refutes his ideas.

If the theory is valid, the division of burdens between C2 systems and the people using them becomes clear. For the C2 system, there are three primary burdens:

- To provide commanders with the information they need to develop and refine command concepts.

- To communicate command concepts—faithfully and clearly—down the chain of command.

- To communicate discovered or impending failures of command concepts—quickly and clearly—up the chain of command.

For the people using the C2 system:

- Command concepts should be primary features of *battle planning*. The consistent mapping of command concepts to combat plans must be made a feature of battle planning at all levels. A spectrum of operationalized command concepts that spans echelons and time should be developed.

- The task of formulating and expressing a command concept should be embedded in *leadership development* and warfighting doctrine.

- Doctrine must ensure that command and control via command concepts is reflected in the manner, content, frequency, and discipline of message traffic on *command* networks.

## DIRECTIONS FOR FUTURE WORK

The research reported here was designed as a preliminary exploration of a general theory of command and control. Our limited objective was to learn whether there was anything obvious in the history of modern warfare that might refute the idea before investing in further research to develop the theory and to apply it to the design and evaluation of C2 systems. There are at least three initial directions for further development of this theory.

The first direction for additional work is to take a different approach to validating the theory: Conduct a series of interviews or discussions with (a) living commanders from all services to reflect their experiences onto our theory and to inquire whether their experiences resonate with or undermine it, and (b) doctrine writers and force developers who are currently grappling with the issue of how to apply technological advances to enhance force effectiveness.

The second direction is to examine the implications for this theory on the real-world problem of the development of C2-system design. In an era of limited resources, what does the theory tell us about how to think about procurement decisions? How do we think about trade-offs between improvements in raw power and enhancements to overall system flexibility? How do we design decision support systems that are empowering but not constraining? What examples from recent history are illustrative of the rewards and pitfalls of making the right or wrong systems decisions?

The third direction is to extend and refine the theory to ensure that it can be generalized over services and their different media (air, land, and sea) for operations. With the growing tendency toward jointness and the blurring of traditional roles and missions between services, it may be that these distinctions are not as important in the present as they may have been in the past. It is also likely that, regardless of medium of combat or service, the pre-battle requirements for information are similar, as is the importance of developing a viable command concept.

# ACKNOWLEDGMENTS

Circumstances—ill health and career changes—prevented the authors from fulfilling their usual roles in report production. Draft fragments of the report languished for several years. The editor, Marian Branch, was the most steadfast advocate for seeing the report through to completion. She took the pieces in hand and resolved to see them melded into a complete report.

Marian undertook the necessary draft writing and editing to complete this work, not just as a chore but as a labor of love. It is her work that appears here every bit as much as that of the principal authors. Under these circumstances, it is common to use the awkward authorship phrase "written with Marian Branch." Instead, we have chosen to honor her great contribution, without which this report simply would not exist, in these paragraphs. She is worthy of being called *an author*.

Marian was greatly assisted by the efforts of artist Sandy Petitjean and proofreader-editor Miriam Polon.

We must also thank Paul Davis for his technical initiative. He saw a need for this report in his reading of the literature. Lastly, Dick Hundley deserves credit for providing management and financial support.

Chapter One

# INTRODUCTION

> If I always appear prepared, it is because before entering on an undertaking, I have meditated for long and have foreseen what may occur. It is not genius which reveals to me suddenly and secretly what I should do in circumstances unexpected by others, it is thought and meditation.
>
> ——Napoleon Bonaparte, 1812[1]

In an age of abundant, almost limitless, information and communications capabilities, decisionmakers are increasingly faced with the problem of too much information, rather than too little. In today's information-oriented society, winnowing, filtering, correlating, and fusing information have become as important as acquiring the information, or (regrettably) even as important as its content, if not more so. Understanding what information is most essential for decisionmaking—so that the information being communicated, processed, or displayed can be bounded—is now a major issue in the design of computer-aided decision support systems.

Nowhere has the problem of overabundant information become more apparent than in military command and control, where the accelerating technologies of communications and computers have flooded commanders at all levels with so much information that they sometimes seem no longer able to digest or comprehend it. The prevailing approach to this problem is to apply still more technology, in the form of computers and software, to sort through, filter, and display the information in ways that will assist the commander in focusing on the "right" information. This approach, of course, assumes that the commander and his responsibilities, circumstances, and

---

[1]The quotations at the beginning of the chapters are taken from Robert Heinl, *Dictionary of Military and Naval Quotations*, Annapolis, Md.: Naval Institute Press, 1966.

decisions are understood well enough for his informational needs to be anticipated.[2] It also assumes that the C2 issues for commanders are acquiring enough information, sorting through it, and then maintaining connectivity with subordinates so that they can be directed.

Excessive reliance on complex command, control, communications, and intelligence (C3I) systems is insidiously dangerous, and counting on the wrong part of command and control (C2)—on a system emphasizing mainly the control rather than the commander—to ensure success in battle can be a prescription for disaster. Yet most current theories of command and control are hierarchical (they represent information as flowing up and down the chain of command) and system-dependent.[3] They envision the commander as using a C2 system to influence events indirectly, at a distance. The commander issues instructions to subordinates, suggestions to commanders of adjacent units, and requests and reports to supporting units and superiors. He develops and maintains a situational awareness of the area of his operations through reports presented by other people or by electronic systems.[4]

Most C2 theories[5] involve information-push processes, in which the design of the system, type of standard messages and their formats, and positioning and capabilities of communication nodes define the type of information available to the commander. Typically, as events in the battle space[6] unfold, descriptive information flows through the hierarchy back to the commander and his staff. As the situation develops, the commander reacts by assessing the situation, developing plans, and issuing orders and reports. According to this view, the

---

[2]In this report, we use *he/his* throughout for clarity, not to imply gender significance.

[3]System-dependent in the sense that the hardware pieces of the system define it. A more detailed discussion of contemporary command and control modeling approaches is given in the next section and in the Appendix.

[4]Thomas P. Coakley, *C3I: Issues of Command and Control*, Washington, D.C.: National Defense University, 1991, pp. 43–52.

[5]See especially the sections "Communications Connectivity" and "Launch Under Attack" models in the Appendix.

[6]Given the three-dimensional nature of modern warfare, and the fact that many future engagements are likely to be fought on media other than dry ground, this term is probably more apt than *battlefield.*

role of the command and control system is to paint a picture for the commander, and the role of the commander is to make highly interactive decisions.

By placing the commander in the position of a processor of inputs and a generator of messages, traditional approaches to modeling command and control create an ideal commander who is, in a way, a prisoner of events: He must react to developments in the battle space rather than anticipate them. A logical corollary is that real-time information becomes critical to the commander's ability to understand and decide. Because the commander makes decisions in reaction to events as they occur and not in anticipation of them, information—lots of it, painting as complete a picture as possible of the battle space—becomes his most critical need.

This approach to C2 clearly views the commander as a reactor, searching the ebb and flow of situational data for critical pieces of information in real time. It proceeds from the notion that the commander is unlikely to anticipate the development of events with any degree of accuracy. It assumes that the commander, his staff, and his supporting C2 system must sift through masses of data—coming in at ever-increasing rates—to glean the relevant clues that will inform his action. It also assumes that the vast majority of information transmitted will be descriptive, and that the most significant cognitive activity of the commander will be "pattern-matching": recognizing the picture and its significance.

The problem is that this approach does not consider the content of what is being transmitted—or what *should* be transmitted. It implies that system designers are able to determine the type and content of the messages that a commander might need and that command and control is simply a function of the hardware, software, and doctrine for its use; if they achieve a master data-fusing system employing the right filters and data-reduction techniques, the commander's decisions can be cued in a timely manner. The system is viewed as a type of magic "bat-signal" that shines in the sky[7] to alert the commander that something important is happening. More significantly,

---

[7]The metaphor is taken from the Batman comic books and films, wherein the police of Gotham City signal their need for the "caped crusader" by shining a searchlight into the sky with a projected image of a stylized bat.

it implies that the information commanders need and will react to is both knowable and invariant for all commanders. In this approach to C2 design, "one size fits all."

Therefore, whereas most contemporary *discussions* of command and control (and, indeed, much of the current military-journal literature on the *practice* of command and control) pay strong lip service to the importance of the human element, there is little in the *theoretical* literature of command and control that does not have the commander boxed up in a wiring diagram. Much of this literature deals with organizations and communications[8] and explains theories with organizational charts that relate commanders to the people and functions they control, or with network-wiring diagrams that relate nodes and links in terms of informational functions or capacities. Defining the C2 process as a function of how the communications system is wiried together is analogous to considering a particular tank gun as being essential to a general theory of ballistics. However, although better guns shoot better, the properties of a gun do not change the fundamental laws of ballistics, which drive the design of the weapon. The rapidly evolving technology of C2 has shifted our attention from the essence of command and control—the individual, idiosyncratic approach of a commander to command that goes beyond military training and doctrine—to its silicon handmaidens, beguiling us with their siren songs of ever more communications, computing, and displays. It seems that a general theory of C2, if one can be determined, should drive the design of C2 systems, good ones of which are absolutely essential for effective performance on the battlefield but cannot substitute for a general theory of command and control.

In this report, we propose an alternative theory of command and control that focuses not on the sufficiency of bandwidth, interoperability, information overload, and stocks, flows, filters, and transformers, but on the cognitive processes of the commander. Specifically, we mean those processes that develop a concept of

---

[8]Two notable exceptions are Martin van Creveld, *Command in War*, Cambridge: Harvard University Press, 1985, especially Chapters 1 and 8; and C. Kenneth Allard, *Command, Control, and the Common Defense*, New Haven: Yale University Press, 1990 (revised 1996). Both of these works are critical of the mechanistic view described above; but they stop short of suggesting a theoretical concept of what C2 *is*. We discuss van Creveld's ideas in Chapter Two.

impending operations that cue the commander (and his C2 system) to look only for certain pieces of information—the substance rather than the means of communications between commanders and their subordinates. To more carefully set our theory apart from the prevailing theories, we briefly describe the most prominent C2 theories here.

## THE FOUNDATIONS OF EXISTING C2 THEORY

To qualify as a theory of command and control, a proposed model must explain widely observed properties and behaviors in terms of more fundamental, or deeper, concepts that draw their principles and vocabularies from analogs in other systems or sciences—control theory, cognitive science, organization theory, neurophysiology, and information theory—not merely describe existing command and control systems. In spite of the apparent diversity of analogs informing these approaches, most models in the literature are fundamentally similar in that they can all be reduced to a variant of a *cybernetic approach*, which describes C2 processes within the framework provided by control theory, or mechanical-electrical communications theory.[9]  Because of this convergence, it is not unreasonable to refer to the dominant approach to modeling command and control as a *cybernetic paradigm*. This is the term we use in the following discussion to describe the standard approach to modeling C2.

### Explaining C2 with Control Theory

The processes of battle—coordinating the activities of multiple independent units and adapting to exogenous changes—are similar to activities encountered in the control of industrial processes. For this reason, control theory provides a powerful framework within which to model the *control* aspect of command and control.

Cybernetic models divide systems into subsystems (components) that exchange signals (inputs and outputs) and that introduce math-

---

[9]Alexander H. Levis and Michael Athans, "The Quest for a C3 Theory: Dreams and Realities," in Stuart E. Johnson and Alexander Levis, eds., *Science of Command and Control: Coping with Uncertainty*, Washington, D.C.: Armed Forces Communications and Electronics Association (AFCEA) International Press, 1988, pp. 4–9.

ematical transformations of those signals. When this approach is applied to C2, it results in models consistent with the cybernetic paradigm. A frequently cited model of this type is that of J. S. Lawson, shown in Figure 1.1.[10] The influence of cybernetics and control theory on Lawson's model is quite clear. Terms such as "desired state" and "sense" are not native to the military lexicon. Indeed, the diagram of Lawson's model could apply equally well to a thermostat as to an industrial control system. Lawson's model is typical of the reactive, picture-painting view of C2 described earlier in this chapter, in which the entire environment provides signals that must be "sensed," evaluated, and compared with a desired state so that their relevance can be determined. In Lawson's world, the commander reacts to signals rather than anticipating them.

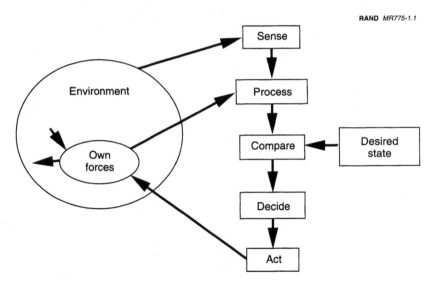

RAND *MR775-1.1*

SOURCE: J. S. Lawson, "Command and Control As a Process," *IEEE Control Systems Magazine*, March 1981, p. 7. Copyright © 1981 IEEE.

**Figure 1.1—Lawson's Model**

---

[10]J. S. Lawson, "Command and Control As a Process," *IEEE Control Systems Magazine*, March 1981, pp. 5–12.

## Organization Charts and C2 Modeling

Organizing military forces into a hierarchy of "units" may be among the most ancient of command and control techniques. Modeling the command and control of any particular armed force certainly requires representing its organizational structure, which encompasses both unit identities and chains of command.

Understanding the functions performed by various units in a military organization can be complex and challenging. However, when viewed as an information-processing mechanism, a military organization operates by exchanging messages (orders, directives, status reports). When flows of information are added to the chart, the result is a model based on flows of information and its transformations at various nodes (Army forces/Joint Force land component, Navy forces/Joint Force maritime component, for example). Figure 1.2 displays an example of such a model (arrows indicate sample nodes).

## MODELING C2 WITH COGNITIVE SCIENCE

Purely cybernetic models—with their numeric signals and transforms—inadequately represent the complex and idiosyncratic activities of humans in C2 systems. To overcome this deficiency, cognitive constructs representing human decisionmaking have been inserted into, or overlaid upon, cybernetic models. Modeling the command part of C2 clearly requires some model of command decisionmaking. Cognitive science provides a rich portfolio of such constructs. A variety of cognitive techniques has been used to model command decisionmaking—in particular, rule-based expert systems and subjective expert utilities with Bayesian updating.[11] Petri nets have been used to model data-flow and decisionmaking structures.[12]

---

[11]Rex V. Brown, "Normative Models for Capturing Tactical Intelligence Knowledge," in Stuart E. Johnson and Alexander H. Levis, eds., *Science of Command and Control: Coping with Complexity*, Fairfax, Va.: AFCEA International Press, 1989, pp. 68–75; Gary A. Klein, "Naturalistic Models of C Decision Making" and Karen L. Ruoff et al., "Situation Assessment Expert Systems for C3I: Models, Methodologies, and Tools," in Johnson and Levis, 1988, pp. 86–92 and 118–126.

[12]D. Tabak and A. H. Levis, "Petri Net Representation of Decision Models," *IEEE Transactions on Systems, Man, and Cybernetics*, Vol. SMC-15, No. 6, 1985.

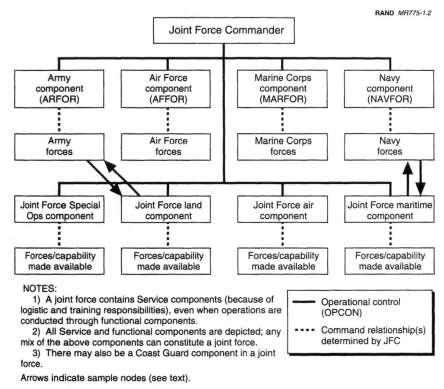

NOTES:
1) A joint force contains Service components (because of logistic and training responsibilities), even when operations are conducted through functional components.
2) All Service and functional components are depicted; any mix of the above components can constitute a joint force.
3) There may also be a Coast Guard component in a joint force.

Arrows indicate sample nodes (see text).

———— Operational control (OPCON)

▪ ▪ ▪ ▪ Command relationship(s) determined by JFC

SOURCE: Adapted from Joint Chiefs of Staff, *Doctrine for Joint Operations*, Washington, D.C.: Office of the Joint Chiefs of Staff, Joint Pub 3-0, February 1995.

**Figure 1.2—Possible Components in a Joint Force**

These compound approaches ultimately rely on chains of command and communications paths for connections between nodes. Therefore, they end up being driven by cybernetic formalisms. Moreover, the human behaviors at all of the nodes are reduced to a common, rational actor: The "human" transformer at one node is the same as that at any other node. The advance, if any, is really one of inserting a more complex mechanical processor at each node.

Thus, regardless of whether the original inspiration comes from control theory or cognitive science, the process of seeking a "deep" theory of command and control produces convergent evolution toward a common destination: a collection of information flows and transforms—boxes and arrows—with the boxes representing processing nodes and the arrows representing information flows. Although cognitive science should be able to generalize the information flows from the real numbers of control theory to arbitrary data structures, the modeling emphasis has remained *not* on the content of the information flow but on the architecture of the boxes and arrows.

In describing C2 systems in this fashion, the content of the information that moves throughout the system and the transformations of that information are secondary to the representation of the nodes and links themselves, and can be represented only in the context of a particular architecture. J. G. Wohl's SHOR (Stimulus-Hypothesis-Option-Response) paradigm, shown in Figure 1.3, is an example. Similar to Lawson's model, this paradigm divides C2 processes into boxes, but its use of such concepts as "hypothesis" indicates that it draws inspiration from cognitive science as well as control theory.[13]

The ability to construct cognitive models that are descriptively accurate is much more poorly established than it is for models drawn from communications or control theory. Still relatively immature, cognitive science provides tools for modeling only certain aspects of command. In particular, the reactive aspect of human decision-making—e.g., picking from a list of preplanned options based on a situation estimate—is much better understood than the leadership aspects of command. Thus, while cognitive science has provided a basis for modeling command beyond the representations used in control theory, the resulting models remain primarily cybernetic in character, neglecting those aspects of command that are not reactive.

---

[13]J. G. Wohl, "Force Management Decision Requirements for Air Force Tactical Command and Control," *IEEE Transactions on Systems, Man, and Cybernetics,* Vol. SMC-11, No. 9, September 1981, pp. 618–639.

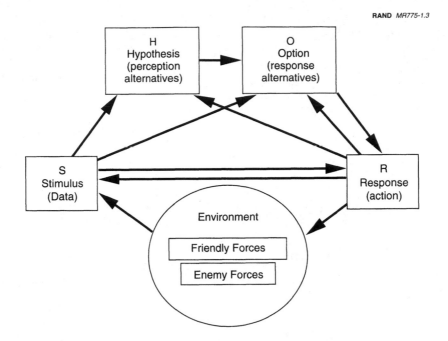

SOURCE: Adapted from J. G. Wohl, "Force Management Decision Requirements for Air Force Tactical Command and Control," *IEEE Transactions on Systems, Man, and Cybernetics*, Vol. SMC-11, No. 9, September 1981, p. 625. Copyright © 1981 IEEE. Adapted by permission.

**Figure 1.3—Wohl's SHOR Model**

Viewing a military organization primarily as an information-processing mechanism neglects many aspects of command but enables C2 models to be constructed without confronting these difficult aspects. Again, this is essentially a cybernetic paradigm. The chief problems with this representation are that the system defines the roles of the humans and that message formats, the type and frequency of messages, and connectivity define what information is available to the commander.

In order to solve the problems of contemporary and future command and control—problems that, for the most part, are unexplained by cybernetic theories of C2—a new theory of command and control that addresses those phenomena neglected by cybernetic models must be considered.  This new theory must explicitly apply across multiple time scales of battle, from readiness and preparation to maximum combat intensity.  It must address the social and cultural aspects of C2, and not reduce commanders to atomized information processors.  Most of all, it must focus on the creativity of commanders.[14]

## A THEORY OF COMMAND CONCEPTS

Thus, cybernetic models are incomplete:  Although they provide a robust basis for understanding *control* functions, they are inadequate to properly descibe *command*, whose human elements cannot be captured in a computer program.  Kenneth Allard, in commenting on the following definitions of *command, command and control,* and *command and control system* in the *Department of Defense Dictionary of Military and Associated Terms,* notes that "one of the most striking characteristics of these definitions is the extent to which they evoke the personal nature of command itself, especially the fact that it is vested in an individual who, being responsible for the 'direction, coordination, and control of military forces,' is then legally and professionally accountable for everything those forces do or fail to do":[15]

> *Command:* "The authority vested in an individual of the armed forces for the direction, coordination, and control of military forces."

> *Command and control:* "The exercise of authority and direction by a properly designated commander over assigned forces in the accomplishment of the mission.  Command and control functions are performed through an arrangement of personnel, equipment, communications, facilities, and procedures which are employed by a commander in planning, directing, coordinating, and controlling forces and operations in the accomplishment of the mission."

---

[14]*Creativity* encompasses a wide range of thought processes and behaviors. The case histories of successful command concepts such as MacArthur's at Inchon and Nimitz's at Midway illustrate individual aspects of creativity.

[15]Allard, 1996 rev., pp. 16–17.

> *Command and control system:* "The facilities, equipment, communications, procedures, and personnel essential to the commander for planning, directing, and controlling operations of assigned forces pursuant to the missions assigned."[16]

When we use these three terms, the above definitions are implied in them.

Going beyond personality alone, our theory suggests that the essence of command lies in the cognitive processes of the commander—not so much the way certain people do think or should think as the ideas that motivate command decisions and serve as the basis for control actions: Ideally, the commander has a prior concept of impending operations that cues him (and his C2 system) to look for certain pieces of information. Rather than seeing the commander and his C2 system as omnivorous consumers of all available information, we see the commander's ideas as generating an information-pull process: a selective searchlight in a sea of information.[17] His technical systems may provide warnings, but they do so primarily because of the contrast of pertinent information against the background of his expectations, which are rooted in a prior, *expressed* concept of operations. If the commander has shared that concept with others, then the warnings can be provided by a strategically positioned subordinate, by a logical conclusion his staff deduces from the presence or absence of a key piece of expected data, or by the commander's own intuitive sense that events are developing either as he had expected or contrary to his expectations.

*Sharing* is the operative word in this concept, and in this way is similar to what FM 100-7 terms the *commander's intent,* which is "a concise expression of the commander's expected outcome of an operation."[18]  But even more than the expression, we focus on the evidence of the cognitive process—the *command concept*—that

---

[16]U.S. Joint Chiefs of Staff, *Department of Defense Dictionary of Military and Associated Terms,* Washington, D.C.: Office of the Joint Chiefs of Staff, JCS Pub. 1, January 1986, p. 74 (quoted in Allard, 1996 rev., p. 16).

[17]Van Creveld calls this the "directed telescope" (1985, p. 75).

[18]Headquarters, Department of the Army, *Decisive Force: The Army in Theater Operations,* Washington, D.C.: FM 100-7, May 31, 1995, p. 5-16.

underlies that expression.  Looking across the history of military operations, from antiquity to the present, and considering the substance rather than the means of communication between commanders and their subordinates, what comes through most consistently is a vision of a military operation—what could and ought to be done in the application of military force against an enemy.  We find renowned commanders mostly concerned with explaining and asking after their vision or expectations of possible and desirable operations:  "Are things going as we planned (envisioned)?  If not, what is broken and needs fixing?  Why and where are things going wrong?  Is the plan (vision) wrong, or does it simply need some adjustment?"

Evidence of command concepts is found most often in war and battle plans, sometimes in the setting of military objectives, less often in the deployment and commitment of forces, and perhaps least often in the issuance of direct orders.  Here, we define a *command concept* as a vision of a prospective military operation that informs command decisions made during that operation.  As such, a command concept may provide an important clue to the minimum essential information that must flow within C2 systems.

Not only should a comprehensive theory of C2 be able to explain how to organize, connect, and process information, it should also

- explain how the quality of commanders' ideas and the expression of those ideas can be assessed and, indeed, duplicated

- explain how C2 systems, including commanders, *should* work, and the *ideal* circumstances in which that work can occur

- provide measures of performance for commanders and their staffs, as well as for the communications and computers that support them.

The motivation behind the theory is a need to separate the *intellectual* performance of the commander from the *technical* performance of the C2 system.  By demonstrating this difference, we demonstrate that the evaluation of C2 systems can finally be separated from the responsibilities of commanders.

Taken to the extreme, the notion of command concepts invites the following hypotheses:

- The most essential functions of command and control are conveying (to subordinates) and altering (for superiors) command concepts.

- All other information in the C2 system is likely to be super-fluous—even detrimental if it diverts attention or effort away from those essential functions.

Ideally, then, battle commanders need only convey their vision of the operation to their subordinates. And the only information subordinates need provide their superiors is what would alter their superior's vision of the operation. In theory, therefore, if a commander's vision of battle was sound and was fully conveyed to subordinates beforehand, there would be no need for information to be in the C2 system during the ensuing battle. Conversely, needing a given amount of information in the C2 system during the battle relates directly to failures associated with the validity or completeness of the command concept or its clear conveyance to subordinates.

*Decisive Force* indicates how the *design,* or concept, enables this lim-iting of information:

> The commander's intent is the central goal and stand-alone refer-ence that enables subordinates to gain the required flexibility in planning and executing. It is the standard reference point from which all present and future subordinates' actions evolve.

> Commanders and leaders—guided by their commander's intent—who can make decisions can better ensure the success of the force as a whole when conditions are vague and confusing and communication is limited or impossible. *The design of com-mander's intent is not to restrain but to empower subordinates by giving them freedom of action to accomplish a mission.*[19]

---

[19]Headquarters, Department of the Army, 1995, p. 5-16. Emphasis added.

## ORGANIZATION OF THE REPORT

In this report, we present a theory of C2 that cuts through the technological overlay[20] that now burdens the subject and attempts to reconcile some familiar instances of C2 success and failure from military history with intellectual performance—with what Napoleon refers to in the quotation at the beginning of this chapter as "thought and meditation." We examine whether empirical patterns can be derived from the structure and content of historical command concepts.

In the next chapter, we describe why we chose historical battles, the criteria we used to select the six battles from military history, and the process we used to synthesize the ideal command concepts that would have been appropriate to those battles. In each of Chapters Three through Eight, we present one case history. In Chapter Nine, we present conclusions and recommendations for further study. The Appendix describes C2 theories in addition to those described earlier in this chapter.

---

[20]The technological overlay is mostly from communications and computers, which are changing at a remarkable pace. Command and control is conducted mostly in and through human minds, which change much more slowly.

# Chapter Two

# THE CONTEXT OF COMMAND AND COMMAND CONCEPTS

The stroke of genius that turns the fate of a battle? I don't believe in it. A battle is a complicated operation, that you prepare laboriously. If the enemy does this, you say to yourself, I will do that. If such and such happens, these are the steps I shall take to meet it. You think out every possible development and decide on the way to deal with the situation created. One of these developments occurs, you put your plan into operation, and everyone says, "What genius . . ." whereas the credit is really due to the labor of preparation.

———Ferdinand Foch, Interview, 1919

## THE ESSENCE OF COMMAND

Martin van Creveld, in his excellent book *Command in War*, describes the essence of command as the ability to deal successfully with uncertainty, to function effectively in the absence of complete information.[1] He stresses that command is both an organizational function and a cognitive function, and that technology, by itself, is not a panacea. Historical success in command has stemmed from a commander's ability to get the most out of his C2 system through structuring, training, and developing his organization to minimize the constraints imposed by the limitations of contemporary technology:[2]

> Far from determining the essence of command, then, communications and information processing technology merely constitute one part of the general environment in which command operates. To allow that part to dictate the structure and functioning of command

---

[1]Van Creveld, 1985, pp. 268–275.

[2]The Guderian case in Chapter Four is an excellent example of such constraint-minimizing organization.

systems, as is sometimes done, is not merely to become the slave of technology but also to lose sight of what command is all about. Furthermore, since any technology is by definition subject to limitations, historical advances in command have often resulted less from any technological superiority that one side had over the other than from the ability to recognize those limitations and to discover ways—improvements in training, doctrine, and organization—of going around them. Instead of confining one's actions to what available technology can do, the point of the exercise is to discover what it cannot do and then proceed to do it nonetheless.[3]

If command concepts are the most essential pieces of information to be conveyed and altered by a command and control system, then the development and quality of those concepts are every bit as important as the ability of systems to inform and communicate them. Therefore, commanders and their visions of operations count, and excellent C2 systems may not be able to overcome a lack of vision or to compensate for a commander's failures. To believe that advanced C2 technology, in the absence of a sound idea controlling its implementation, will automatically be a "force multiplier" may be turning a blind eye to the historical evidence of both the importance and variability—idiosyncrasy—of commanders and their ideas.

We examine six historical cases that suggest, as did van Creveld, that the quality of the commander's ideas is a critical factor in the functioning of C2 systems. An idea, or command concept, that outpaces the system's ability to support it can cause the C2 system to function poorly and the command function to suffer. Likewise, if the concept is flawed and does not identify the information critical to validating it, the system will likely become flooded with extraneous information as units in battle attempt to make sense out of what is happening. The cases also suggest that the massive improvements in C2-system performance over the past several decades have not altered the reality that people have to know what to look for in order to maximize the performance of the system.

---

[3]Van Creveld, 1985, p. 275.

## SELECTION CRITERIA

In selecting the six historical cases, we were, of course, looking for information that would give insights into the C2 theory advanced here. More specifically, we selected the cases according to the following criteria:

- The conflicts are modern (they occurred in the past 50 years), because modern warfare involves a scale of operations and technology that can affect C2 systems and their use.

- They represent different battle media (land, sea, and air). Each medium may present different C2 challenges and solutions.

- The account in the literature appeared likely to be adequate for us to discern the presence or absence of command concepts.

We limited the cases to six so that, within our resource constraints, we could examine them in the detail necessary to illuminate the theory. We selected the particular cases because we were interested personally in learning more about them as military history.

The six cases span four modern wars and eight principal combatants. The combatants appear in more than one case, in both offensive and defensive roles:

- World War II, with the Germans on the offensive against the French at Sedan, the Japanese on the offensive against the Americans at Midway, and the British (principally) on the offensive against the Germans at MARKET-GARDEN.

- The Korean Conflict, with the Americans on the offensive against the North Koreans at Inchon.

- The War in Vietnam, with the Americans on the offensive against the North Vietnamese.

- The Gulf War, with the Americans (principally) on the offensive against the Iraqis.

They also represent the spectrum of operational media and forces:

- naval, air, and amphibious forces and operations at Midway and Inchon

- air and mechanized forces in blitzkrieg operations at Sedan and in DESERT STORM

- airborne forces and ground operations at MARKET-GARDEN and Ia Drang.

And the command concepts in those cases run the gamut from those that were

- visionary and determinedly pursued, as at Midway and Inchon

- to well-developed and fully articulated, as at Sedan and in DESERT STORM

- to fundamentally flawed and incomplete, as at Ia Drang and MARKET-GARDEN.

We do not impute any significance to the correlation between the last two lists—the quality of the command concepts and the dominant kinds of forces involved (naval, mechanized, and airborne); the cases are too few and we can imagine too many exceptions. We also focus on operational or strategic commanders because battle accounts often focus on their role, and their quotations reveal their thought processes.

## ELEMENTS OF THE COMMAND CONCEPT

If it could be demonstrated that command concepts—their promulgation and correction—are (or should be) the essential substantive content of the information flowing through C2 systems, then the following approach to a theory of command and control is indicated: The structure and content of historical command concepts—actual or implied—should be examined for empirical patterns, particularly as those patterns may vary by command environment, circumstances, and level.

An intellectual device for carrying out the examination is to synthesize the ideal command concepts that would have been appropriate to some well-documented battles.[4] Even in those (many) cases in

---

[4]Admiral Horatio Nelson's writings before Trafalgar and the history of the battle itself may provide a good example of what is pertinent to a near-ideal command concept.

which the commander did not have a clearly articulated command concept or was not successful in battle, it should be possible to prepare an ideal command concept to match the actual events of the battle.

An ideal command concept for a historical battle is a hypothetical statement of the commander's intent that should have been, under the doctrine, training, and common knowledge of the time, clearly sufficient for subordinate commanders to successfully execute the responsibilities they were actually called on to fulfill during battle, without exchanging additional information with their superior commander.  In effect, it answers the question, "What would the commander have had to tell his subordinates before the battle in order to have made their subsequent actions conform to his concept?"

Translation of concept to information is the key step:  If conveying and altering concepts are the essence of ideal C2 *functions*, then conveying information to those same ends is the priority task of C2 *systems*.

Our theory proposes the command concept as the primary feature of *battle planning*.  At a minimum, the command concept should include the following elements:

* Time scales that reveal adequate preparation and readiness not just of the concept but of the armed forces tasked with carrying it out.  Ideally, there would be enough time to take what van Creveld considers the optimal approach to improving the performance of a C2 system:  Divide the task into various parts and create forces capable of dealing with these parts on a semi-independent basis.[5]  A commander who develops his command concept on the fly, based primarily on a picture of the battle space, even in real time, has probably already lost his fight, because he has failed to develop a mental picture of probable future developments in the battle space.

* Awareness of the key physical, geographical, and meteorological features of the battle space—situational awareness—that will enable the concept to be realized.

---

[5]Van Creveld, 1985, p. 269.

- Congruence between the plan and the means (resources, troops) for carrying it out.

- A structuring of forces consistent with the battle tasks to be accomplished. Several of the case histories involve all the services in a way that supports the forces of the one service tasked with the strategic mission. MARKET-GARDEN illustrates the problems of trying to make a task conform to the available service/forces.

- What is to be accomplished, from the highest to the lowest levels of command. The squad leader understands his platoon commander's concept for platoon operations and has then developed his own for squad operations that is supportive of and consistent with the next-higher-level concept. The role of doctrine is to make such nested, hierarchical concepts consistent.

- Intelligence on what the enemy is expected to do, including the confirming and refuting signs to be looked for throughout the coming engagement.

- What the enemy is trying to accomplish, not just what his capabilities and dispositions may be.

- What the concept-originating commander and his forces should be able to do and how to do it, with all of its problems and opportunities—not just the required deployments, logistics, and schedules, but the nature of the clashes and what to expect in the confusion of battle. Personality or character traits of key officers are sometimes mentioned because they are viewed in historical accounts as contributing significantly to how an operation unfolds.

- Indicators of the failure of or flaws in the command concepts and ways of identifying and communicating information that would change or cancel the concept. Stresses on the C2 system are included among these indicators so that we can examine whether the manner in which different commanders develop, articulate, and execute their ideas places different burdens on their C2 systems.

- A contingency plan in the event of failure of the concept and the resulting operation.

This list parallels the statement of the operational commander's intent set forth in FM 100-7:

> After mission analysis, the operational-level commander clearly describes the operation's purpose, the desired end state, the degree of acceptable risk, and the method of unifying focus for all subordinate elements. The operational-level commander's intent contains the intent statement of the next senior commander in the chain of command. The commander's intent is meant to be a constant reference point for subordinates to discipline their efforts. It helps them focus on what they have to do to achieve success, even under changed conditions when plans and concepts no longer apply. For major operations, a clear statement of intent is essential to successful integration and synchronization of effort, including support operations throughout the depth of the battle space.[6]

In some—but not all—of the historical cases that follow, the commander has provided an operational concept that reveals these elements. For this reason, we have structured the histories so that the concepts can be inferred. Each chapter is organized as follows:

• A background section describing the physical/geographic situation, key players, intelligence sources, and whether those sources were relied on

• The operational plan

• A description of the battle

• A description of the communications, or command and control

• A command concept drawn from secondary-source descriptions of the above elements. The concept is divided into three sections: information about the enemy and his plans, information about the concept-originating commander and his plans, and contingency plans. We provide quotations from the secondary sources to reveal the thinking-out-loud of the commander, insights into the creativity/idiosyncrasies of the commander that may refute doctrine or training, and the demands placed on the C2 system.

---

[6]Headquarters, Department of the Army, 1995, p. 1-2.

- Our assessment of how well the battle adheres to our theory of what is absolutely essential information for a C2 system.

Note that although the concepts are presented in a linear format, the cognitive processes themselves may not have been linear.

History provides many more examples of situations in which the command concept was neither sound nor even formed, neither shared nor agreed upon, and neither altered in the face of contrary information nor susceptible to being altered. These less-happy examples do not refute the notion of command concepts or their centrality to command and control. Rather, they emphasize the importance of command concepts and demonstrate that there are many command and control failures that even perfect C2 systems cannot prevent.

After we present the case histories, we summarize them in Chapter Nine. We focus on those elements that were "looked for" and found and/or those that were not looked for but should have been. Because of the differences among battles, an element that is crucial to one battle may be minor or barely mentioned in regard to another.

We begin with a case history that bears out our theory most clearly in the realization of the commander's vision: Nimitz at the Battle of Midway.

# Chapter Three

# MASTER OF THE GAME: NIMITZ AT MIDWAY

> The most complete and happy victory is this: to compel one's enemy to give up his purpose, while suffering no harm to oneself.
>
> ——Belisarius, 505–565

## BACKGROUND

Following a series of naval defeats in the Pacific that began with the disastrous attack on Pearl Harbor and ended with a draw in the Battle of Coral Sea, the United States won its first decisive naval battle of World War II at Midway in early June 1942. The Japanese planned to attack and occupy Midway Island at the extreme end of the Hawaiian Island chain,[1] extending their perimeter in the Pacific[2] and hoping to draw the U.S. naval forces into a decisive battle.[3] The attack and occupation were to be carried out by three major Japanese naval forces steaming independently toward Midway (see Figure 3.1):[4]

- a carrier strike force centered on four aircraft carriers approaching from the northwest[5]

---

[1]Samuel Elliot Morison, *History of United States Naval Operations in World War II, Vol. 4, Coral Sea, Midway and Submarine Actions, May 1942–August 1942*, Boston: Little, Brown and Company, 1949, p. 70.

[2]Morison, 1949, p. 74.

[3]Morison, 1949, p. 75.

[4]Morison, 1949, pp. 87–93, provides the order of battle for these three Japanese forces, as well as that for the U.S. forces available for the defense of Midway.

[5]Also called the first mobile force, under the command of Vice Admiral Chuichi Nagumo (Morison, 1949, p. 88).

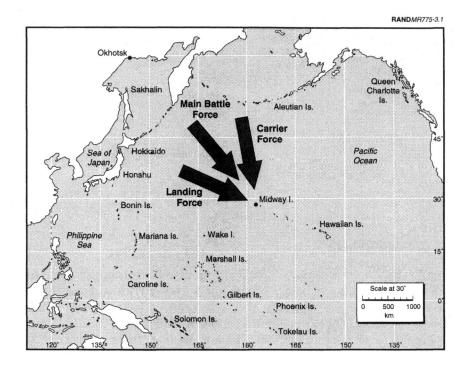

**Figure 3.1—The Japanese Concept of Operation**

- an invasion or occupation force with transports approaching
  from the west[6]

- a main battle force on battleships approaching between the first
  two, so as to support both.[7]

A diversionary attack by the Japanese was to be made on the Aleutian
Islands at the same time. The larger plan was as follows:

---

[6]Also called the Midway occupation force, under the command of Vice Admiral
Nobutake Kondo (Morison, 1949, p. 88).

[7]Also called the main body and the First Fleet, under the command of Admiral Isoroku
Yamamoto (Morison, 1949, p. 89).

Admiral Yamamoto [was] determined to complete the neu-
tralization of Pearl Harbor.  His weapon would again be the
airplane, but his objective was twofold.  The entire Combined Fleet
would accompany an invasion force to Midway Island in the
western Hawaiian group.  Presumably the Americans would commit
their remaining two or three carriers to defend Midway.  Nagumo's
four carriers or Yamamoto's battleships, far superior to the enemy,
would destroy the unescorted US carriers.  The Japanese Army
would then land on Midway, capture it, and convert the airfield into
a base for . . . the systematic bombing of Hawaii.  With the carriers
sunk and Pearl Harbor under constant air attack, the United States
might then realize the fruitlessness of trying to fight in the Pacific.[8]

The first objective of the Japanese carrier strike force was to "execute
an aerial attack on Midway . . . destroying all enemy air forces
stationed there"[9] in preparation for the intended landing of the
invasion or occupation force.

United States Navy cryptographers in Hawaii, working for Admiral
Chester W. Nimitz, Commander in Chief of all Allied armed forces in
the Pacific Ocean area (CINCPAC), were able to read fragments of the
coded Japanese naval communications.[10]  By correlating these
fragments with intelligence on Japanese forces and operations, they
were able to deduce the outlines of the Japanese plans.  But

> Washington remained skeptical.  For one thing, they still hadn't
> pinned down exactly what the Japanese meant by "AF."  Rochefort
> [cryptography] was always sure it was Midway but he needed proof.
> Around May 10 he went to Layton [intelligence] with an idea.  Could
> Midway be instructed to radio a fake message in plain English,
> saying their fresh-water machinery had broken down?  Nimitz
> cheerfully went along with the ruse . . . Midway followed through . . .

---

[8]Clark G. Reynolds, *The Fast Carriers: The Forging of an Air Navy*, Annapolis, Md.:
Naval Institute Press, 1968, p. 11.

[9]Morison, 1949, p. 95.

[10]E. B. Potter, *Nimitz*, Annapolis, Md.: Naval Institute Press, 1976, p. 64, describes the
code as "roughly 45,000 five-digit groups . . . most of which represented words and
phrases.  As a means of frustrating cryptanalysis, a book of 50,000 random five-digit
groups was issued to Japanese communicators.  The sender added a series of these
random groups to the code groups of his message. . . . To further foil cryptanalysis, the
Japanese from time to time issued new random-group books . . . ."

and two days later a Japanese intercept was picked up, reporting that AF was low on fresh water.[11]

Still, some were not convinced that the enemy would attack Midway and not Hawaii proper or even the West Coast of the United States. When Nimitz briefed Major General Delos Emmons, the local Army commander in Hawaii, Emmons pointed out that the intelligence was predicated on intent, not capabilities; and the Japanese possessed the capability to attack Hawaii. Nimitz did not back away from his staff's estimates, but, being both careful and conciliatory,

> he assigned one of his staff, Captain J. M. Steele, to the specific job of keeping an eye on the Combat Intelligence Unit's material. Steele became a sort of "devil's advocate," deliberately challenging every estimate, deliberately making Rochefort and Layton back up every point.[12]

Despite General Emmons reservations, Nimitz had already "made the first vital decision of the campaign in accepting the estimate of his fleet intelligence officer that Midway and the Aleutians were the enemy's real objectives":[13]

> As early as 20 May [two weeks before the attack] Admiral Nimitz issued an estimate of the enemy force that was accurate as far as it went—and even alarming. . . . Although the picture was not complete, the composition, approximate routes and timetable of the enemy forces that immediately threatened Midway were so accurately deduced that on 23 May, Rear Admiral Bellinger, the Naval air commander at Pearl, was able to predict the Japanese plan of attack. . . .[14]

By May 25, a little more than a week before the attack, further details of the Japanese plans were laid bare, including "the various units, the ships, the captains, the course, the launching time—everything . . .

---

[11]Walter Lord, *Incredible Victory*, New York: Harper & Row, 1967, p. 23.

[12]Lord, 1967, p. 25.

[13]Morison, 1949, p. 80.

[14]Morison, 1949, p. 80.

the exact battle order and operating plan of the Japanese Striking Force."[15]

Midway was reinforced against the impending Japanese attack, but neither the quantity nor the quality of the forces that could be stationed there was sufficient to give high confidence of a successful defense. Search plans were drawn up to detect the approaching Japanese fleets. It was at this point that the possibility emerged of a flanking attack on the Japanese carrier strike force that was expected to approach Midway from the northwest. Nimitz's staff aviation officer noted that "the plan will leave an excellent flanking area northeast of Midway for our carriers."[16]

The naval strike forces available in Hawaii were limited to three aircraft carriers, one of them recently damaged in the Battle of Coral Sea. Admiral Nimitz dispatched Admiral Raymond Ames Spruance[17] with two of the carriers as Task Force 16 and, subsequently, Admiral F. J. Fletcher with the third, quickly repaired carrier as Task Force 17 to protect Midway and to inflict damage on the Japanese forces:

> Admiral Fletcher, as senior to Admiral Spruance, became O.T.C. (Officer in Tactical Command) as soon as their rendezvous was effected. As he possessed no aviation staff . . . it was probably fortunate that Spruance exercised practically an independent command during the crucial actions. . . . Neither [Spruance] nor Fletcher exercised any control of the air and ground forces on Midway Island, over the submarines deployed in their area, or over [the] force in the Aleutians. The overall commander was Admiral Nimitz, who remained perforce at his Pearl Harbor headquarters.[18]

---

[15]Lord, 1967, p. 27.

[16]The staff officer was Captain A. C. Davis (Lord, 1967, p. 29).

[17]Of Nimitz's choice of Admiral Spruance, Morison (1949, p. 82) has this to say: "A happy choice indeed, for Spruance was not merely competent; he had the level head and cool judgment that would be required to deal with new contingencies and a fluid situation; a man secure within."

[18]Morison, 1949, p. 85.

By the terminology of that time, Admiral Nimitz in Hawaii retained broad tactical control of all forces engaged in the defense of Midway.[19]  By today's terminology, we would say that Admiral Fletcher had tactical control and Admiral Nimitz had operational control.[20]

## THE PLANS

On the basis of the intelligence available to him, Admiral Nimitz drew up an operational plan (CINCPAC Operation Plan No. 29-42) for defending Midway.  This plan was remarkably detailed with respect to the Japanese fleet elements that were involved, their lines of approach, and timing.  "This catalog of chilling details was followed by the American answer:  an outline of the tactics [Nimitz] proposed to follow.  Specific tasks were assigned each of the various US forces."[21]

Although some of the details were reasonable conjectures about Japanese naval practices, many were the result of intelligence gathered or correlated through cryptographic analysis of Japanese naval communications.  Events would prove the plan to be accurate in all of its essentials.  The plan was made available to all subordinates who were directing forces, on Midway, afloat, and even those who would take to the air:

> To those eligible to see it, the meticulous intelligence on the Japanese movements seemed almost incredible.  Not knowing where it came from—and perhaps having read too many spy thrillers—[one of the carrier officers] could only say to himself, "That man of ours in Tokyo is worth every cent we pay him."[22]

---

[19]Morison (1949, p. 79) puts it this way: "Admiral Nimitz . . . exercised strategic and *broad tactical direction* of all American forces, naval or military, deployed in the Pacific Ocean. . . ." Potter (1976, p. 89) puts it another way: "Admiral Nimitz, acting as coordinator, was retaining *overall tactical command*—land, sea, and air." (All emphasis added.)

[20]RADM James A. Winnefeld, USN (Ret.), in a comment on an earlier draft of this case study.

[21]Lord, 1967, p. 35.

[22]Lord, 1967, p. 35.

According to plan, the three U.S. carriers were deployed to a position northeast of Midway where they could bring their aircraft to cover Midway as well as to bear on the flank of the Japanese carrier forces that were expected to approach Midway from the northwest (see Figure 3.2). They were less well positioned to reach the Japanese invasion and main battle forces, which were expected to approach Midway from more westerly directions:

> [Spruance] saw at once the advantage of placing his force on [the Japanese strike force's] flank and the possibility of attacking the Japanese carriers while their planes were raiding Midway. Spruance also made the prudent observation that the US carrier forces should not proceed west of Midway in search of the enemy before the enemy carriers were substantially disabled. The Japanese might

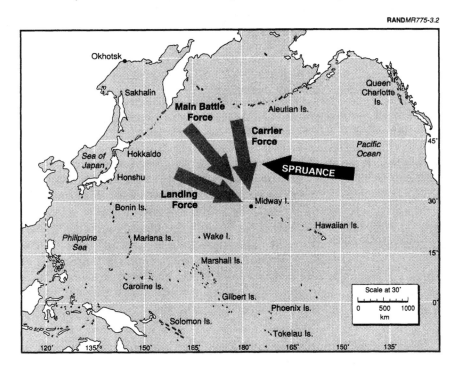

Figure 3.2—Nimitz's Concept

alter their plans and head for Pearl Harbor, in which event the American forces might find themselves bypassed and unable to intervene.[23]

[Spruance] had already made certain definite decisions. He would not come within 700 miles of Wake Island, no matter what the temptation. He knew the Japanese had beefed up the place, and he did not want to mix with land-based aviation. Nor did he intend to permit the Japanese to draw him so far west that they could close in with their superior surface strength and clobber him.[24]

The U.S. carrier forces were directed to protect Midway by finding and attacking the Japanese carriers. In pursuit of those objectives, they were generally directed to balance the risk to the U.S. carriers against the damage that might be inflicted on the Japanese:

In carrying out the task assigned . . . you will be governed by the principle of calculated risk, which you shall interpret to mean the avoidance of exposure of your force to attack by superior enemy forces without good prospect of inflicting, as a result of such exposure, greater damage on the enemy.[25]

## THE BATTLE

The three Japanese fleets approached Midway as anticipated by Nimitz's intelligence and operational plan. On June 3, 1942, with the Japanese carrier strike force to the northwest of Midway still undetected under the cover of seasonal fog and overcast skies, the U.S. aircraft based at Midway discovered the Japanese invasion fleet approaching from the west. The message sent into Midway from a Navy flying boat said, "Main body . . . bearing 262 [almost due west of Midway], distance 700 [miles]. . . ."[26] Upon hearing this, Nimitz took the precaution of relaying the message to his carrier forces in the event they had not heard it directly themselves and then added the

---

[23]Potter, 1976, pp. 84, 85.

[24]Gordon W. Prange, *Miracle at Midway*, New York: McGraw-Hill Book Company, 1982, p. 112.

[25]Morison, 1949, p. 84; Lord, 1967, p. 36.

[26]Potter, 1976, p. 91.

warning, "That is not repeat not the enemy striking force—stop—That is the landing force. The striking force will hit from the northwest at daylight tomorrow."[27] Army bombers based at Midway then attacked the invasion fleet, but with only modest success.[28]

The main show was expected the following day, on the morning of June 4:

> It was assumed that the Japanese Striking Force would begin launching at dawn—attack planes southward toward Midway, search planes north, east, and south. At that hour the American [carrier] task forces, on course southwest through the night, should be 200 miles north of Midway, ready to launch on receiving the first report from US search planes of the locations, course, and speed of the enemy. With good timing and good luck they would catch the Japanese carriers with half their planes away attacking Midway. With better timing and better luck they might catch the enemy carriers while they were recovering the Midway attack group. That the Americans might catch the Japanese carriers in the highly vulnerable state of rearming and refueling the recovered planes was almost too much to hope for.[29]

> At dawn, June 4 . . . the report they were awaiting . . . came, an urgent message in plain language, sent via the cable from Midway: "Plane reports two carriers and Main Body ships bearing 320, course 135, speed 25, distance 180."[30]

Upon hearing this, Nimitz remarked to his intelligence officer (Layton), "Well, you were only five miles, five degrees, and five minutes off."[31] By this time, the Japanese carrier air strike was on its

---

[27]Prange, 1982, p. 170.

[28]They were able to damage one Japanese tanker (Morison, 1949, p. 99).

[29]Potter, 1976, p. 87.

[30]Potter, 1976, p. 93

[31]Potter, 1976, p. 93. Prange (1987, p. 102) quotes Layton as having said, "I anticipate that first contact will be made by our search planes out of Midway at 0600 Midway time, 4 June, 325 degrees northwest at a distance of 175 miles." Potter, 1976, p. 83, quotes Layton slightly differently when Nimitz pressed him to be specific: "All right then, I've previously given you the intelligence that the carriers will probably attack Midway on the morning of the 4th of June, so we'll pick the 4th of June for the day. They'll come in from the northwest on bearing 325 degrees and they will be sighted at about 175 miles from Midway, and the time will be about 0600 Midway time."

way to Midway.  The Midway-based aircraft had been launched at dawn for another attack on the invasion fleet to the west, but were diverted against the Japanese carrier strike force as soon as it revealed itself to the northwest.  However, the results were even less than those against the invasion force on the previous day:

> [The] land-based air attacks on the morning of 4 June resulted only in severe losses to the Midway-based groups and some dearly-bought experience.  An examination of Japanese records and the interrogations of Japanese officers by United States Army officers after the war will convince the most optimistic that no damage and only a few casualties were inflicted on enemy ships by land-based planes, whether Army, Navy or Marine Corps, on the Fourth of June.[32]

That indictment, however, does not speak to the effect of these attacks upon the Japanese.  The attacks from the Midway-based aircraft made the Japanese realize that the air threat from Midway had not been eliminated and that they would have to reattack the island with their carrier aircraft.  It was this threat of further air attacks from Midway that led to the undoing of the Japanese carriers:[33]

> The Midway air force was not, in practice, to inflict crippling damage on Yamamoto's fleet; but its existence was to invest American capabilities with a psychological menace that troubled the Japanese throughout the battle, while its intervention, on one if not two occasions, was to distort their decision-making with disastrous effect.[34]

Meanwhile, the U.S. carriers, lurking in "the fog of war" on the flanks to the northeast, were listening to the radio reports between Midway

---

[32]Morison, 1949, p. 111.

[33]Immediately after the first Japanese carrier air strike against Midway, the Japanese air commander radioed back to the carriers, "There is need for a second attack wave." (Morison, 1949, p. 107).  One of the authors (CHB) is indebted to Rear Admiral James Winnefeld, USN (Ret.), for bringing this point about the value of the land-based air attacks to his attention.

[34]John Keegan, *The Price of Admiralty:  The Evolution of Naval Warfare*, New York: Penguin Books, 1990, p. 225.

and its aircraft and prepared to launch their aircraft against the Japanese carriers:

> Spruance had originally intended to launch his planes . . . when there would be less than a hundred miles of ocean for them to cover, provided the Japanese carriers maintained course toward the atoll. But, as reports came in of the strike on Midway, he decided, on the advice of his chief of staff, to launch two hours earlier in the hope of catching the carriers in the act of refueling planes on deck for a second strike on the atoll.[35]

The U.S.-carrier-based torpedo bombers arrived first and sustained heavy losses without inflicting any damage on the Japanese carrier force.[36] Again, however, the effect of the attack was to be measured not so much in damage to Japanese ships as to Japanese decisions.

The dive-bombers had difficulty in locating the carrier strike force and arrived late, but they caught the Japanese at an extreme disadvantage:

*   The Japanese had successfully repelled waves of attacks from the U.S. Midway-based bombers and then from the carrier-based torpedo bombers.

*   Their fighter protection had been drawn down to low altitudes while engaging the U.S. torpedo bombers.

*   They had recovered their aircraft from the first strike on Midway and were equivocating between arming them for another attack on Midway or a strike against the U.S. carriers, which had now made their presence obvious.

Without prior coordination, the dive-bombers from the three U.S. carriers separately and successfully attacked three of the four Japanese carriers, wrecking them and leaving them sinking. Aircraft from the fourth Japanese carrier then found and attacked the single carrier of Task Force 17 under Admiral Fletcher's command, damaging it but leaving it salvageable and its aircraft able to land on

---

[35]Morison, 1949, p. 113.

[36]Morison points out that "out of 41 torpedo planes from the three carriers, only six returned, and not a single torpedo reached the enemy ships" (1949, p. 121).

the two carriers of Task Force 16 under the command of Admiral Spruance. After this, Admiral Fletcher effectively relinquished the U.S. command afloat to Admiral Spruance.

Admiral Spruance launched a second strike in search of the Japanese carriers, found the fourth one as well as the three burning wrecks, and completed the destruction of all four Japanese carriers. He then retired to the east during the night in order to avoid a night surface engagement with the Japanese main battle force:

> Although not yet sure that [the fourth Japanese carrier] had sunk, Spruance found that knocking off the last Japanese carrier left him with a problem: *What do we do now?* he asked himself. *If we steam westward what will we run into during the night?* He had no use for futile heroics, and did not care to invite a night engagement with his carriers. The Japanese had far superior surface forces, they excelled in night operations, and Spruance did not have enough destroyers to screen and protect his flattops.[37]

Indeed, the Japanese main battle force under Admiral Yamamoto initially planned to pursue the American carriers for just that opportunity; however, "Yamamoto realized that, if he persisted in pushing forces eastward in search of a night battle, he was likely to get a dawn air attack instead."[38] Reluctantly, Yamamoto ordered a general retirement of the Japanese forces.

Early the next day, Spruance turned westward again and conducted further air strikes against the remaining Japanese fleets, which were now mostly retreating to the west. In his report of the action, Spruance said,

> I did not feel justified in risking a night encounter with possibly superior enemy forces, but on the other hand I did not want to be too far away from Midway the next morning. I wished to have a position from which either to follow up retreating enemy forces or to break up a landing attack on Midway.[39]

---

[37]Prange, 1987, p. 301. Emphasis as in the original.

[38]Morison, 1949, p. 139.

[39]Morison, 1949, p. 142, citing CTF 16 Action Report, June 16, p. 3.

Prange defends Spruance's caution, noting that "his mission was to protect Midway. This he could not do by leaving it to the tender mercies of a Japanese invasion force. . . ." Moreover,

> the possibility of Task Forces Sixteen and Seventeen annihilating the [entire] enemy fleet had never crossed Nimitz's mind or anyone else's. In canceling out four Japanese carriers, Spruance and his men had done all and infinitely more than was expected of them.[40]

Although other air, submarine, and surface actions were involved in the Battle of Midway, the carrier strikes were the decisive engagements. Three U.S. carriers had destroyed four Japanese carriers with the loss (ultimately, to a Japanese submarine) of only one of their own.[41] The Japanese plan to invade Midway could not be completed because air superiority had been lost to Midway's defenders. The battle is generally considered to have been the turning point of the war in the Pacific.

## COMMAND AND CONTROL

For the most part, Spruance and Fletcher faithfully executed Nimitz's operational plan. The only discretion they exercised under that plan was deciding precisely when to launch their air strikes and when to retire or advance from the flanking position. The communications between Spruance and Fletcher prior to the battle were limited to visual signals, because of their fear of being detected through radio transmissions. Even so, the following visually transmitted message from Spruance to the ships of his Task Force 16 two days before the battle gives some idea of the clarity of the impending situation to the tactical commanders:

> An attack for the purpose of capturing Midway is expected. The attacking force may be composed of all combatant types including four or five carriers, transports and train vessels. If presence of Task Forces 16 and 17 remains unknown to enemy we should be able to make surprise flank attacks on enemy carriers from position northeast of Midway. Further operations will be based on result of

---

[40]Prange, 1987, p. 302.

[41]Prange, 1987, p. 396, provides a tally of the losses on both sides during the battle.

these attacks, damage inflicted by Midway forces, and information of enemy movements. The successful conclusion of the operation now commencing will be of great value to our country. Should carriers become separated during attacks by enemy aircraft, they will endeavor to remain within visual touch.[42]

Once Spruance and Fletcher sailed from Hawaii, Nimitz had no closed command-and-control loop (two-way communications) with them; he could broadcast to them, but they could not return his calls without revealing their positions. As noted earlier in this chapter, when aircraft from Midway detected the Japanese invasion force to the west and incorrectly reported it by radio as "the main body," Nimitz, fearing that his carrier task forces might be misled, broadcast a correction, hoping his carrier forces were listening. Otherwise, Nimitz's knowledge of his forces afloat was based solely on his own plan and his assumption that it was being followed.

On the eve of the battle between the carriers, Nimitz broadcast a message to his carrier task force commanders that was not unlike the exhortation to duty in the signal hoisted by Nelson as his fleet went into battle at Trafalgar:

The situation is developing as expected. Carriers, our most important objective, should soon be located. Tomorrow may be the day you can give them the works. The whole course of the war in the Pacific may hinge on the developments of the next two or three days.[43]

Communications between Midway and Hawaii were not so circumscribed, because a submarine cable between the two was not subject to enemy monitoring or exploitation:

Another ace up Nimitz's sleeve was the Pacific cable which since 1903 had linked Honolulu to Manila, with Midway as one of its stations. This undersea line carried the bulk of the heavy pre-battle communications exchange between Pearl Harbor and Midway—a line which the Japanese could not tap. Normal radio traffic from

---

[42]Morison, 1949, p. 98, and E. P. Forrestal, *Admiral Raymond A. Spruance, USN: A Study in Command*, Washington, D.C.: Department of the Navy, 1966, pp. 39, 40.

[43]Potter, 1976, p. 92.

shore to ship thus could not give the Japanese a true picture of what the Americans were up to.[44]

Moreover, unlike the carriers, radio communications from both Midway and Hawaii could not be exploited by the enemy to locate the U.S. forces; the positions of the islands were certainly well known. So, Nimitz watched the battle unfold mostly from what he heard from Midway or overheard from the sporadic reports of aviators and submariners as they detected or engaged the enemy. But for the queen on his chessboard, his three carriers, he had to rely upon his plan and upon the men to whom he had entrusted its execution.

## NIMITZ'S COMMAND CONCEPT

Recall that an ideal command concept for an historical battle is a hypothetical statement that, under the circumstances prevailing at the time, would have been clearly sufficient for subordinate commanders to execute successfully the responsibilities that actually befell them during the battle, without exchanging additional information with their superior commander. For the Battle of Midway, something close to the ideal appears to have existed in writing: Nimitz's Op Plan 29-42. It may be that an ideal command concept for Midway would have been even sparer than that plan.

Indeed, the available literature suggests that an ideal command concept could be written for Midway quite easily from scratch now, without prior inspection of Op Plan 29-42. It might consist of the following:

### I. ABOUT THE ENEMY AND HIS PLANS:

1. The enemy [Japan] is expected to attack Midway on the morning of June 4 for the purposes of (a) occupying Midway and (b) engaging and defeating our [U.S.] naval forces in a decisive battle.

---

[44]Prange, 1987, p. 105.

2. The enemy attack will be carried out by three forces: (a) a carrier strike force approaching Midway from the northwest, (b) an occupation force approaching from the west, and (c) a main battle force approaching from a line between these two. All these forces will be accompanied by heavy screens of cruisers and destroyers and will be preceded by submarine and patrol aircraft operations.

3. You should expect the attack to begin with an enemy carrier air attack upon Midway for the purpose of suppressing its air defenses prior to its being bombarded by surface ships and occupied.

4. If, at any time, the enemy should detect major elements of our naval forces within range, you should expect that he will divert his forces to effect their engagement and destruction.

## II. ABOUT OUR FORCE DISPOSITIONS AND PLANS:

1. We shall reinforce Midway with the purposes of (a) defending the island against occupation, (b) inflicting damage on the attacking enemy forces, and (c) locating the enemy forces so that other U.S. forces can be brought to bear.

2. We shall deploy all available carrier forces to the north and east of Midway so that they may be brought to bear (a) on the flank of the enemy carrier strike force, and (b) in support of the defense of Midway.

3. Our strategic objective is the defense of Midway, but our tactical objective is to inflict as much damage as possible upon the enemy carriers with the least exposure and damage to our own. The measure of tactical success will be disproportionate damage to the Japanese carriers.

4. If the enemy carriers can be put out of action while we retain our air operations from either Midway or our carriers, the enemy cannot prevail in his objectives. Therefore, the primary enemy targets for all of our aviation strike assets, at Midway and on our carriers, should be the enemy carriers.

5. To protect our carriers from exposure, (a) maintain absolute radio silence until their position has been unambiguously ascertained by the enemy, and (b) withhold air strikes until the enemy carriers have been located.

## III. ABOUT CONTINGENCIES:

1. We should take care not to expose our carriers to superior enemy forces in the forms of (a) night surface actions, (b) battleship gunfire, and (b) land-based aviation from Wake Island.

2. Our aircraft are more replaceable than are our pilots; and both are more replaceable than our carriers.  If the enemy carriers are located, press the attacks without regard to the recovery of our aircraft.

3. Be prepared, before launching air strikes against the enemy carriers, to distribute or balance the weight of your attacks among the enemy carriers that may present themselves as targets.

4. Locating the enemy naval forces before the enemy locates ours will be crucial to our success.  Therefore, deploy submarine and aircraft patrols aggressively into the sectors where the enemy is expected.

## ASSESSMENT

The Battle of Midway would appear to be a near-perfect manifestation of our command concepts theory:

- A commander (Nimitz) with a sound and detailed concept (thanks to outstanding cryptographic and intelligence analysis) of how to succeed in an impending battle.

- The thorough communication of that concept to subordinates through a detailed operational plan (86 copies were made of Op Plan 29-42).[45]

- The need for little in the way of command and control communications during the execution of that concept, either up (advising of errors in the concept) or down (corrective adjustments in the concept).

The only departure from the ideal was the commander's intervention when he felt compelled to correct the interpretation of an aviator's sighting report about the location of the enemy's main force. The commander could as well have said, "Ignore that report, which is at variance with my concept; stick to my concept!" As it turned out, even that communication proved unnecessary: His subordinates (Spruance and Fletcher) in control of the critical forces (the carriers) knew that the sighting report was incorrect and had every intention of following through with the commander's concept. Nimitz could have gone on extended vacation when his carriers left Pearl Harbor; his concept was sufficient to carry the burden of battle and ensure the victory.

Spruance insisted after the war that "the credit must be given to Nimitz. Not only did he accept the intelligence picture but he acted upon it at once."[46] And those actions include the dissemination of a clear concept to all who needed it so that they could turn it into a reality.

Whereas much was left to the discretion of commanders hand-picked by Nimitz, the concept that is the subject of Chapter Four demonstrates a thoroughly developed planning cycle—five months spent in rehearsal according to doctrine—and a drive conducted counter to the expectations of the enemy.

---

[45]Lord, 1967, p. 35. According to Prange (1982, p. 99), the plan was distributed to "all task force, squadron, and division commanders."

[46]Prange, 1982, p. 393.

# THE TECHNICIAN: GUDERIAN'S BREAKTHROUGH AT SEDAN

Get there fustest with the mostest men. . . .

—Nathan Bedford Forrest

## BACKGROUND

From May 10 through May 15, 1940, General Heinz Guderian's XIX Panzer Corps left their assembly areas on the western border of Germany, broke through the north end of the Maginot Line, and turned a decisive victory at Sedan into a rout of the entire French Army. The offensive ended the so-called phony war, during which most of the German Army, flush with the victory in Poland, moved into assembly areas along the western border of Germany and conducted training and preparation for an offensive operation into France.

From mid-October through late December 1939, the German General Staff produced a series of plans for the campaign, almost all of which bore a strong similarity to the von Schlieffen Plan used by the Germans in 1914. The Allies guessed that the Germans would use such a plan, and although the planning staffs on both sides were heartily unenthusiastic about the prospect of another trench war in France, preparations proceeded apace.

## THE PLANS

The German General Staff plan, code-named *FALL GELB* (CASE YELLOW), consisted of two Army Groups, A and B, attacking on a wide front from Metz to Venlo, with Army Group B (commanded by Field Marshal Fedor von Bock) in the north, designated the main effort. Its mission was to conduct a rapid wheeling movement through the Low Countries while Army Group A (commanded by Colonel-General Gerd von Rundstedt) tied up most of the French Army along the Franco-German border. For these purposes, Army Group B was given most of the motorized forces, since it would have to move very fast to succeed; Army Group A inherited the majority of the foot-mobile and horse-drawn formations.[1]

After winter war games had uncovered a potential opportunity for a swift advance through the Ardennes, General Fritz von Manstein raised the issue with Adolf Hitler. On February 18, 1940, Hitler approved a bold stroke through the Ardennes with a fast and powerful tank and motorized infantry force.[2] To accomplish this task, he allocated an additional six motorized divisions to von Rundstedt's Army Group under the command of General Ewald von Kleist. The force was designated *Panzergruppe von Kleist* and was given the mission of punching a narrow salient through the Ardennes with the objective of forcing a crossing of the Meuse River at Sedan. The spearhead of *Panzergruppe von Kleist* was XIX Panzer Corps, a four-corps formation consisting of five panzer divisions, four motorized infantry divisions, and a flak corps. It was commanded by General Heinz Guderian.[3] Figure 4.1 shows the new plan, as modified by Hitler.

*Panzergruppe von Kleist* had the task of traversing some of the most difficult terrain in Europe, which required that six major rivers be crossed over a five-day period. Given the limited trafficability and narrowness of the avenues of approach in the sector, von Kleist's commanders developed a number of tactical innovations in applying motorization to maneuver and in achieving cooperation between arms of service. Using infantry, armor, engineers, artillery, and air support as a combined-arms team, the panzer units were able to move and fight with a rapidity that was, at the time, breathtaking.

---

[1]T. Dodson Stamps and Vincent J. Esposito, eds., *A Military History of World War II*, West Point: U.S. Military Academy, 1953, Vol. I, pp. 66–70; B. H. Liddell Hart, *History of the Second World War*, New York: G. P. Putnam's Sons, 1970, p. 65.

[2]Eliot Cohen and John Gooch, *Military Misfortunes*, New York: Free Press, 1990, p. 202.

[3]Liddell Hart, 1970, p. 65.

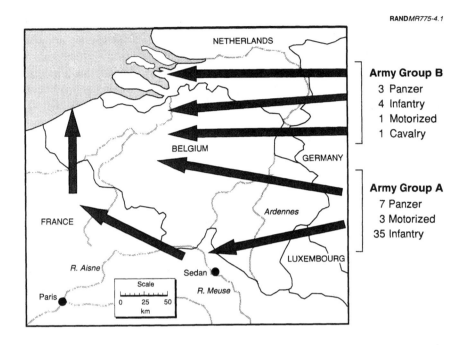

RAND*MR775-4.1*

Figure 4.1—*FALL GELB* (CASE YELLOW) Modified Operational Concept

## THE CAMPAIGN

On May 10, 1940, at 0535 hours, XIX Panzer Corps initiated the offensive. Advancing in three columns through the Ardennes (2d Panzer Division in the north, 1st Panzer Division in the center, and 10th Panzer Division in the south), the corps found its main difficulty to be traffic control rather than enemy action. Despite the initial confusion, it advanced through Luxembourg on schedule.[4] On the evening of May 10, the lead elements of the 1st Panzer Division reached the obstacles that marked the Belgian border and began work in earnest to clear the obstructions.[5]

---

[4]Stamps and Esposito, 1953, p. 75.

[5]Stamps and Esposito, 1953, p. 75.

To gain a firsthand impression of the progress of his corps, Guderian spent most of the day with forward elements of his divisions. His order for the following day, May 11, consisted mainly of a reiteration of his intent to secure the west bank of the Meuse. Guderian's intent for the 1st Panzer Division for May 11 was to break through the Belgian second line of resistance at Neufchateau and, if possible, reach the west bank of the Meuse at Sedan.[6]

Extremely difficult terrain and stiffening French and Belgian resistance frustrated the achievement of that goal. However, by day's end, the lead elements of the 1st Panzer Division were 5 kilometers from the French border and 20 kilometers from Sedan. Guderian's concept for the operation remained the same, and he issued an order to that effect just before midnight on May 11.[7]

On May 12, the 1st Panzer Division covered the remaining 20 kilometers to the banks of the Meuse in just 4 hours, and by day's end, the entire corps had closed on the east bank of the Meuse just opposite Sedan. Although the XIX Panzer Corps had spent nearly five months rehearsing this operation, Guderian insisted on ensuring that his intent was understood:

> General Guderian spent the entire morning [of May 13] visiting his three division commanders, conducting face-to-face coordination, and explaining his aims for the upcoming operation.[8]

The French had not yet panicked. They assumed that the Germans would advance in a manner based on the French experience with logistics. That is, the Germans would have to stop at the Meuse, the first river line defended by a major system of fortifications. There, the Germans would have to consolidate and prepare for the river crossing for several days, perhaps as long as a week. General Maurice-Gustave Gamelin, the French commander, ordered an additional 11 divisions to reinforce the Sedan area. They would arrive during May 14–21. Guderian understood that success depended on the speed with which the panzer forces could get across the Meuse

---

[6]Julian K. Rothbrust, *Guderian's XIX Panzer Corps*, London: Cassell, 1974, p. 60.

[7]Rothbrust, 1974, p. 61.

[8]Rothbrust, 1974, p. 66.

and into the open country on the western side—before the French could identify this threat and take steps to neutralize it. On May 12, von Kleist approved Guderian's request to attempt a crossing without waiting for heavy infantry reinforcements to arrive.[9]

Guderian's XIX Panzer Corps crossed the Meuse on the fly, straight from the march. After an intense bombardment of the river defenses by the Luftwaffe, lasting nearly the entire day, elements of the 1st Panzer Division, led by Lieutenant-Colonel Hermann Balck's 1st Panzer Grenadier Regiment, managed to cross and establish a toehold on the west bank of the Meuse at Glaire, just north of Sedan, during the early evening hours of May 14 (see Figure 4.2):

> The soldiers of the 1st Panzer Division, main effort of the Panzer Group von Kleist, observed the devastating Luftwaffe attack the entire day. Nevertheless, the situation was chaotic when they began their river crossing, with French bunkers spitting intense fire at them from the far side of the river . . . at 1500 hours, under the protection of the massive air attack and subsequent artillery preparation, infantry and engineers carried their boats to the water's edge . . . [Balck's boats arrived without operators] . . . Balck had trained his soldiers in the use of pneumatic boats, thus he decided to conduct the assault crossing without the help of engineers. He crossed the river with the first wave and within minutes reached the initial bunker line along the far bank. The advance slowly began to increase momentum. . . . By midnight, Balck had led elements of his regiment to just south of Cheveuges and to the southern edge of the Bois de Marfee.[10]

During the night of May 13, Guderian's engineers managed to erect a bridge across the Meuse, and Guderian pushed more than 150 armored vehicles across the bridge that night. When morning came, he had a coherent force on the far bank and began to push for a breakout.

---

[9]Liddell Hart, 1970, p. 71.

[10]Rothbrust, 1974, p. 74.

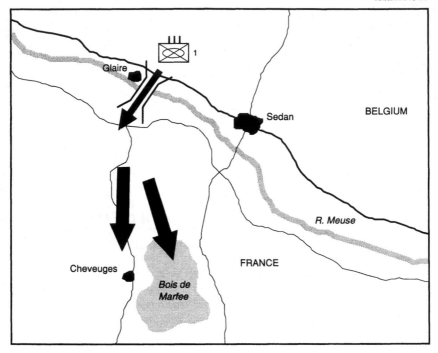

**Figure 4.2—The Meuse Crossing**

Von Kleist began insisting on consolidation to destroy remaining French forces, to remove a potential threat to his flanks. According to Guderian:

> "An order came from Panzer Group Headquarters to halt the advance and confine the troops to the bridgehead gained. I would not and could not put up with this order, as it meant forfeiting surprise and all our initial success."[11]

---

[11]B. H. Liddell Hart, *The Other Side of the Hill,* London: Cassell, 1951, p. 177.

After a lively argument with von Kleist, on the telephone, the latter agreed "to permit the continuation of the advance for another twenty-four hours—in order to widen the bridgehead."[12]

For the next five days, Guderian struggled with higher headquarters, because he was continually ordered to halt and consolidate his gains. Through a variety of artifices, including receiving permission to conduct "strong reconnaissance" to the west, Guderian's advance guard had, on May 19, arrived in Abbeville, on the Channel coast.[13] The French armies had not been destroyed, but Guderian's lightning advance had destroyed their center of gravity, and their leadership had collapsed. Large numbers of French units had surrendered with their fighting capability intact. Guderian's XIX Panzer Corps had achieved in nine days what the German Army in 1914 had failed to achieve in four years.[14]

## COMMAND AND CONTROL

Guderian was a consummate opportunist—not in a pejorative sense, but in his understanding of the effect of speed and surprise on his opponent's operational center of gravity. Of the crossing, the French commander, General Gamelin, commented after the war:

> It was a remarkable maneuver. But had it been entirely foreseen in advance? I do not believe it—any more than Napoleon had foreseen the maneuver at Jena, or Moltke that of Sedan [in 1870]. It was a perfect utilization of circumstances. It showed troops and a command that knew how to maneuver, who were organized to operate quickly—as tanks, aircraft, and wireless permitted them to do. It is perhaps the first time that a battle had been won, which became decisive, without having had to engage the bulk of the forces.[15]

Guderian's command and control system, consisting of a mixture of wire, short-range tactical radio, and runners, could not inform him of the appearance of opportunities in sufficient time to exploit them.

---

[12]Liddell Hart, 1970, p. 72.

[13]Liddell Hart, 1970, p. 72.

[14]Rothbrust, 1974, p. 88.

[15]Liddell Hart, 1951, p. 181.

His response was to physically position himself where he could sense the pulse of the battle and make and rapidly disseminate key decisions. When he *was* absent, he depended on his subordinate commanders' and staffs' understanding of his intent to keep their units operating in a manner that supported his overall goals. This understanding of the commander's intent, and consequent granting of wide latitude of action to subordinates within the boundaries of that intent—a principles-based doctrine—was a feature of *Wehrmacht* operations.[16]

> During offensive operations, adherence to fundamental rules, doctrinal concepts, and a solid plan significantly contributed to the German Army's success. . . . Actions at operational and tactical levels resulted from commanders clearly understanding von Brauchitz's [Chief of German General Staff] intent.[17]

Not only intent but a clear understanding of time-and-space relationships and the capabilities of other types of units were drilled into the panzer formations down to a very low level of command:

> Commanders down to battalion level in Guderian's Panzer divisions understood the operational concept in 1940, and thus were able to take full advantage of unexpected circumstances. Intense training during winter and spring, at all levels, prepared commanders for the mental challenges of making those critical decisions. The numerous war games conducted at army group, army, and corps provided them with the opportunity to study all aspects of the upcoming battle. At the unit level, repeated river crossings at locations closely resembling the actual crossing sites allowed leaders to rehearse their tasks until they became second nature. . . . Through this rigorous training period, leaders perfected the mobile warfare doctrine that ultimately led them to victory in France.[18]

---

[16]This operational methodology, known as *Auftragstaktik* (loosely translated as "mission-oriented operation"), is currently in vogue in the U.S. Army and frequently appears in doctrinal articles as an example to emulate.

[17]Rothbrust, 1974, p. 92.

[18]Rothbrust, 1974, p. 93.

## GUDERIAN'S COMMAND CONCEPT

At least three command concepts informed the German operations on the Western Front:  (1) the *FALL GELB* plan of the German General Staff (a variation on the von Schlieffen Plan), (2) von Manstein's overlaid concept of punching through the Ardennes, and (3) Guderian's improvised concept of how best to exploit the breakthrough at Sedan by heading directly to the coast.  All three were realized to some degree, but it was Guderian's concept riding on the back of von Manstein's that produced the triumph of the blitzkrieg in 1940.  Up to the breakthrough at Sedan, Guderian executed von Manstein's command concept; thereafter, his concept dominated the operation—even over the opposition of his superiors.

From the planning and execution of this operation, it is evident that the professional development of the German commanders, at least at an operational level, was extraordinarily high.  The technology embedded in the panzer formations (high mobility, good tactical communications) fully supported the exploitation of this command style.  The Germans' understanding of "how to do" this modern style of war and of the significant advantages offered by personal leadership— backed by superior staff organization—allowed the German commanders to dominate their French counterparts:

> Through years of "efficiency-aimed" training and a common doctrine, [German] staffs were able to dispense with lengthy operation orders during the actual campaign and simply operated on fragmentary orders.  The concept of commanders at the front insured more face-to-face discussion between commanders and subordinates, contributing not only to higher confidence levels in command, but also furnishing a clear understanding of the leader's aims.[19]

For this reason, the XIX Panzer Corps' advance to the Meuse River in 1940 was a textbook application of a well-articulated and executed command concept. After the crossing of the Meuse, Guderian modified that concept to exploit an opportunity to fulfill the larger, strategic goals of that concept:  to neutralize the enemy armies.  This

---

[19]Rothbrust, 1974, p. 94.

composite concept, idealized with hindsight and recast in our format, might be stated as follows:

## I. ABOUT THE ENEMY AND HIS PLANS:

1. The enemy [France] currently has four infantry and three tank divisions in our [Germany's] sector. He expects our main effort to be a thrust through the Low Countries.

2. Little enemy resistance is expected until the crossing of the Semois River at Bouillon. Beyond that point, the enemy may be expected to resist fiercely any breakthrough attempt. If the main line of the French resistance at Sedan is penetrated, however, the French ability to resist will collapse.

3. We should expect the French to attempt to reinforce and/or assist the troops garrisoning the fortress areas near Sedan with at most two infantry divisions and 250 to 300 armored vehicles.

## II. ABOUT OUR FORCE DISPOSITIONS AND PLANS:

1. Our strategic objective is the military defeat of France and the moral collapse of the French Army. Our objective is not the physical destruction of the French Army. Our tactical objectives are to (a) breach the Maginot Line, (b) outflank the French frontier defenses, and (c) establish a salient across the Meuse to enable a drive to the Channel coast.

2. The XIX Panzer Corps shall advance through Luxembourg, Belgium, and France with three panzer divisions abreast to force crossings of the Our, Semois, and Meuse Rivers. XIX Panzer Corps will be the main effort of Panzer Group von Kleist. 1st Panzer Division is the corps' main effort.

3. We shall make intensive use of tactical aerial bombardment to isolate the fortress troops at Bouillon and at Sedan to facilitate the advance of the XIX Panzer Corps.

4. Once across the Meuse at Sedan, XIX Panzer Corps will not consolidate the crossing but will immediately seek opportunities for a breakthrough beyond the Meuse.

5. Given a breakthrough beyond the Meuse, XIX Panzer Corps should turn west and drive for the Channel coast, destroying the ability of the Allied armies to control and maneuver their forces.

## III. ABOUT CONTINGENCIES:

1. We must not permit the Belgians to significantly interfere with our offensive operations, despite the defensive advantages the terrain in the Ardennes may afford them.

2. The six water obstacles in-sector must be crossed quickly, without hesitation.

3. XIX Panzer Corps must keep its forward elements supplied and supported by fire from the air while denying the French ability to resupply their fortress troops.

4. The XIX Panzer Corps must penetrate and bypass at Sedan before the French can significantly reinforce that strongpoint.

5. Once across the Meuse, we must seek the best opportunity to neutralize the French Army as a fighting force.

6. The key to our success will be maintaining our forward momentum and advancing faster than the French can react. If XIX Panzer Corps cannot maintain the speed of attack, the French will be able to reorient themselves along interior lines and our corps can become bogged down and be defeated in detail. Therefore, priority of air and artillery fires will be to XIX Corps units successfully advancing in-sector, with the objective of facilitating their advance.

## ASSESSMENT

Guderian's concept was daring, appropriate, and visionary. Like Nimitz's concept at Midway, Guderian's concept was embedded in extensive planning and preparation for the campaign, and was so well internalized by his subordinates that Guderian rarely talked to his division commanders during operations. While Spruance at Midway made a wise *tactical* decision to withdraw his carriers to the east during the night, Guderian made an important *strategic* decision—one contrary to the operational concept of his superior, von Kleist—to exploit his breakthrough by turning west to the English Channel, dividing and demoralizing his enemies.

Unlike the air-sea warfare at the battle of Midway, the nature of ground combat at the time required Guderian to be forward, at the *Schwerpunkt* ("center of gravity"), to exert his personal leadership ability, and to sense when his next trigger point had been reached. It was there, rather than at his headquarters, that Guderian could maintain a situational awareness appropriate to his needs. When Guderian was absent from a particular location, his concept allowed subordinates great freedom of action within the confines of his intent, or, as FM 100-7 maintains, "not to restrain but to empower subordinates by giving them freedom of action to accomplish a mission."[20]   The fact that Guderian required little in the way of communication with his direct subordinates during operations speaks volumes about the quality of his plans.

The German C2 system was just robust enough to accommodate Guderian's leadership style. Often visiting his headquarters for only a few brief minutes each day, he depended heavily on his chief of staff to execute the enormous amount of coordination and planning necessary to keep the iron machines advancing, and in Guderian's case, to keep the General Staff *un*informed of the extent of his progress. Through the lens of our theory, Guderian's campaign offers an example of a leader who, in van Creveld's words, "recognized the limitations of his C2 system and . . . discovered ways— improvements in training, doctrine, and organization—of going around them."[21]

The next case study is in many ways the opposite of this scenario. The C2 system is substantially more advanced than Guderian's— which was just enough to keep up—and the commander, unlike Guderian, is not physically present in the battle space. But the change in concept is similar, as is the degree of success.

---

[20]Headquarters, Department of the Army, 1995, p. 5-16.
[21]Van Creveld, 1985, p. 275.

# Chapter Five

# TECHNOLOGY'S CHILD: SCHWARZKOPF AND OPERATION DESERT STORM

> The blow, whenever struck, to be successful, must be
> sudden and heavy.
>
> ———Robert E. Lee

## BACKGROUND

Operation DESERT STORM was a military episode embedded in a much wider military-political campaign. Waged by Iraq, the campaign aimed at gaining political and economic hegemony over the Persian Gulf region. The military phase, which was to evolve into Operation DESERT STORM, began on July 16, 1990, when a Defense Intelligence Agency analyst noted that a brigade of the Hammurabi Division of Iraq's Republican Guard had moved into southern Iraq, opposite its northern border with Kuwait.[1]

Over the next two days, three divisions, the Hammurabi, Medina Luminous, and In God We Trust, were spotted moving into the same area.[2] During the next week, Iraq moved an additional five divisions to assembly areas close to the Kuwait border. On August 1, these units uncoiled from their assembly areas and deployed in assault formation on the border with Kuwait.

That same day, at 5 p.m. Washington time, Iraq invaded. Initially, two Republican Guard divisions—the Hammurabi and the Medina Luminous—spearheaded the assault. Within three hours, the Iraqi armored forces were in Kuwait City, assisted by a Special Forces di-

---

[1] Michael R. Gordon and Bernard L. Trainor, *The Generals' War: The Inside Story of the Conflict in the Gulf,* Boston: Little, Brown and Company, 1995, p. 4.

[2] Gordon and Trainor, 1995, p. 31.

vision that had been airlifted into the city itself.[3]  As Iraqi armored units piled up on the roads in and around Kuwait City, the ultimate intentions of the Iraqi leadership remained unclear.  By midday on August 3, however, the Iraqis had sorted themselves out and were clearly moving south, threatening the security of Saudi Arabia.

As the situation was developing during the run-up to the invasion, the Chairman of the Joint Chiefs of Staff, General Colin Powell, had instructed the Commander in Chief of Central Command (CINCCENTCOM), General H. Norman Schwarzkopf, to prepare a two-tiered set of military options to respond to a potential invasion: a set of defensive options to protect Saudi Arabia, and a set of offensive options to take the war to the Iraqis if necessary.  On August 4, Schwarzkopf briefed President George Bush on CENTCOM's contingency war plan, OPLAN 90-1002 (or "ten-oh-two"), which laid down the force requirements for *defending* Saudi Arabia, and the transport, logistics, and time required to get them there.  It envisioned a 17-week deployment of three Army divisions, two Marine Expeditionary Forces, and three carrier battle groups to the Persian Gulf—over 200,000 soldiers in all.[4]

Ten-oh-two was briefed to several groups of policymakers.  Finally, President Bush instructed Defense Secretary Dick Cheney to discuss with King Fahd of Saudi Arabia the idea that a massive influx of U.S. military power might be necessary to secure the kingdom.  After several days of negotiations, the Saudis agreed to allow the U.S. military to deploy to the Saudi kingdom to deter an attack on Saudi Arabia.  At 4 p.m. in Washington on August 7, General Powell received the authorization to execute OPLAN 90-1002.  Immediately, two squadrons (48 aircraft) of F-15 fighters and the Division Ready Brigade of the 82d Airborne Division (2,300 soldiers) deployed to Saudi Arabia, arriving at Dhahran on August 8.  Two carrier battle groups arrived on-station at the same time.  By mid-September, the air deployment was almost complete, with over 700 aircraft in place.[5]  Over the following 14 weeks, an additional 230,000 troops arrived.  On December 1,

---

[3]Lawrence Freedman and Ephraim Karsh, *The Gulf Conflict, 1990–1991: Diplomacy and War in the New World Order*, Princeton: Princeton University Press, 1993, p. 67.

[4]Gordon and Trainor, 1995, p. 46.

[5]Freedman and Karsh, 1993, p. 94.

General Schwarzkopf reported to the President that he had accomplished his mission. A force adequate to deter an Iraqi invasion of Saudi Arabia was in place.

Halfway through this process, policymakers in Washington pressed General Powell to develop an offensive option—to take the war to the Iraqis and push them out of Kuwait. On October 10 and 11, Schwarzkopf's deputy, Marine Major General Robert B. Johnston, briefed a rough outline of an offensive plan to the Joint Chiefs and to President Bush.[6] General Powell telephoned Schwarzkopf and asked him what he would need in resources for the offensive option. Schwarzkopf asked Powell for two more ground divisions—specifically, the U.S. VII Corps, which at that time was stationed in Germany.

On October 31, at a White House meeting, General Powell presented Schwarzkopf's "shopping list" to President Bush, along with the caution that it would take an additional three months to get the troops in place for an offensive operation. President Bush agreed to provide the forces necessary for the task, and, on November 1, issued the order.[7] Powell gave Schwarzkopf all he had asked for, and more. He ordered the services in response to send *three* extra Army divisions, a second Marine division, two more carrier battle groups (for a total of six), and over 300 more Air Force aircraft—approximately twice the force that was currently in-country.[8]

At the same time, the United States was successfully marshaling a coalition of nations to support actions against Iraq with the aims of (1) garnering world (and United Nations) approval, (2) keeping the conflict from being seen as an American or Western versus Arab or Islamic conflict, and (3) preventing Israeli involvement, a lightning rod for Arab hostility. The resulting political coalition and the appearance of a military coalition were essential to the success of Operation DESERT STORM. However, despite the significant military participation by other nations, military operations were domi-

---

[6]Gordon and Trainor, 1995, pp. 129, 132–134.

[7]Gordon and Trainor, 1995, pp. 153–154.

[8]H. Norman Schwarzkopf and Peter Petre, *It Doesn't Take a Hero*, New York: Bantam, 1992, p. 376.

nated by U.S. forces and their logistics support. Thus, the descriptions that follow are centered on the U.S. coalition commander and U.S. military operations.

## THE PLANS

Prior to the invasion, Iraq had three publicly stated goals. The first was a demand for adjustments to the Kuwait border in favor of Iraq. The second was forgiveness of Iraq's $40 billion war debt with Kuwait. The third was cession to Iraq of control of the islands of Warba and Bubiyan, which controlled the approaches to the mouth of the Euphrates River, and thus to Iraq.[9] After the invasion, annexation of Kuwait as Iraq's "nineteenth province" was added.

Iraq's strategic objective was to terminate any hostilities so that she would be holding more than she had started with—territory, wealth, and prestige. Iraq's operational objectives were to hang onto as much Kuwaiti territory as possible, by inflicting sufficient punishment on the coalition forces to cause them to sue for peace:

> A strategy of intensive defense, conceding no ground without a hard fight, was Saddam's best hope of achieving his political objective of holding on to as much of Kuwait as possible. The higher the costs imposed, the more the enemy would be prepared to accept a peace on terms that were unobtainable prior to hostilities. . . . [T]o this end, Iraqi forces barricaded themselves into Kuwait. They constructed a massive defensive line close to the border with Saudi Arabia, a mixture of obstacles designed to stop a tank offensive, with coastal defenses prepared to repulse an amphibious assault.[10]

The Iraqis had 1,127 aircraft; a ground force of 900,000 soldiers in 63 divisions, 8 of which were of the Republican Guard; 5,747 tanks, 1,072 of which were modern T-72s; 10,000 armored fighting vehicles, including 1,600 modern BMPs (Soviet armored personnel carriers); and 3,500 artillery tubes. With this force at their disposal, the Iraqi Armed Forces' General Command calculated that the United States

---

[9]Gordon and Trainor, 1995, p. 27.

[10]Freedman and Karsh, 1993, p. 278.

would need a force three times as large—or 3 million soldiers—to throw it out of Kuwait.[11]

A great deal of uncertainty was expressed among U.S. decision-makers about the level of risk involved in an offensive campaign. Many voices were raised, warning of huge numbers of potential casualties. A key question that surfaced repeatedly was whether waging an air-only campaign would reduce risks to U.S. forces. General Powell was steadfast in his adamant opposition to that idea:

> Many experts, amateurs, and others in this town, believe that this can be accomplished by such things as surgical air strikes or perhaps a sustained air strike. And there are a variety of other nice, tidy, alleged low-cost, incremental, may-work options that are floated around with great regularity all over this town [but] one can hunker down, one can dig in, one can disperse to ride-out such a single-dimension attack. . . . Such strategies are designed to hope to win, they are not designed to win.[12]

The fundamental flaw, he argued, was to leave the initiative with the Iraqi president:

> He makes the decision whether he will or will not withdraw. He decides whether he has been punished enough so that it is now necessary for him to reverse his direction and take a new political tack.[13]

By the end of October, President Bush had become firmly convinced that an offensive to drive the Iraqis out of Kuwait was necessary.[14] To accomplish this objectives, the CENTCOM planners came up with a four-phase plan:

---

[11]Freedman and Karsh, 1993, pp. 279–280. This belief was apparently based on the military rule of thumb that a head-on attack into a prepared defense requires a preponderance of three times the defending force for the attacker to prevail—and even then at great cost.

[12]General Powell's testimony before the Senate Armed Services Committee, December 3, 1991, cited in Bob Woodward, *The Commanders*, New York: Pocket Books, 1992, p. 329.

[13]Freedman and Karsh, 1993, p. 286.

[14]Gordon and Trainor, 1995, p. 153.

- Phase 1 was to last between seven and ten days, and was intended to achieve air supremacy and incapacitate Iraq's command and control system. This phase was to be executed by strategic air activity.

- Phase 2 was intended to last several additional days, and was aimed at Iraq's warmaking ability—primarily at weapons of mass destruction (nuclear, biological, and chemical), the eight Republican Guard divisions, and 12 major petrochemical facilities, including three refineries.

- Phase 3 was intended to be an intense bombing campaign in Kuwait proper, aimed at disrupting, demoralizing, and destroying as many of the 400,000 troops occupying Kuwait as possible.

- Phase 4 was a ground campaign designed to surround, isolate, and defeat in detail the Iraqi occupation forces in Kuwait:[15]

> Since Saddam had most of his forces in southern Kuwait and along the Gulf coast to the east, the ground plan called for moving VII Corps several hundred miles in a wide arc to the west, and attacking through Iraq to hit the Republican Guard. It would amount to a gigantic left hook. Massive, swift, crushing tank attacks were central to the plan. . . . The idea was to force Saddam to move his hundreds of thousands of troops from dug-in positions so they could be picked-off with superior US air and ground fire.[16]

After Schwarzkopf was informed of which units he would have at his disposal for an offensive operation, he assembled their commanders in Dhahran on November 14 to brief them on his intent:

> The first thing we're going to have to do is, I don't like to use the word 'decapitate,' so I think I'll use the word 'attack,' leadership, and go after his command and control. Number two, we've got to gain and maintain air superiority. Number three, we've got to totally cut his supply lines. We also need to destroy his chemical, biological, and nuclear capability. And finally, all you tankers, listen to this. We need to destroy—not attack, not damage, not surround—I want you to *destroy* the Republican Guard. When

---

[15]Freedman and Karsh, 1993, p. 301.

[16]Schwarzkopf and Petre, 1992, p. 352.

you're done with them, I don't want them to be an effective fighting force anymore.    I don't want them to exist as a military organization.[17]

Schwarzkopf then continued by explaining *how* CENTCOM would accomplish this task (see Figure 5.1):

> I anticipated, I said, a four-pronged ground assault.   Along the Saudi-Kuwaiti border near the gulf, I wanted two divisions of US Marines and a Saudi task force to thrust straight into Kuwait, with the objective of tying up Saddam's forces and eventually encircling Kuwait City. . . . I'd reserved a second corridor, in the western part of Kuwait, for a parallel attack by the pan-Arab forces led by two armored divisions from Egypt and another Saudi task force.   Their objective would be the road junction north of Kuwait City that controlled the Iraqi supply lines. . . . I indicated a section of Saudi-Iraqi border more than three hundred and fifty miles inland. . . . I wanted Luck's [XVIII Airborne Corps] divisions to race north from that area to the Euphrates, blocking the Republican Guard's last route of retreat. . . . Finally, I turned to Fred Franks [Commander VII Corps]. "I think it's pretty obvious what your mission's going to be," I said, moving my hand across the desert corridor just to the west of Kuwait, "attack through here and destroy the Republican Guard."[18]

For deception, Schwarzkopf instructed XVIII Airborne Corps and VII Corps to maintain their forces in assembly areas near Kuwait, to keep Iraqi forces focused on those two avenues of approach.   As soon as the air war began, the Iraqis would be pinned down and both corps would shift laterally several hundred miles to the west without interference.[19]

On December 29, Defense Secretary Cheney signed the Warning Order to implement DESERT STORM, with a target date for the initiation of the air campaign of January 15, 1991.   At 1030 hours on January 15, President Bush met with his advisers to discuss the text of a National Security Directive (NSD) authorizing the execution of

---

[17]Schwarzkopf and Petre, 1992, p. 381.

[18]Schwarzkopf and Petre, 1992, pp. 382–383.

[19]Schwarzkopf and Petre, 1992, p. 383.

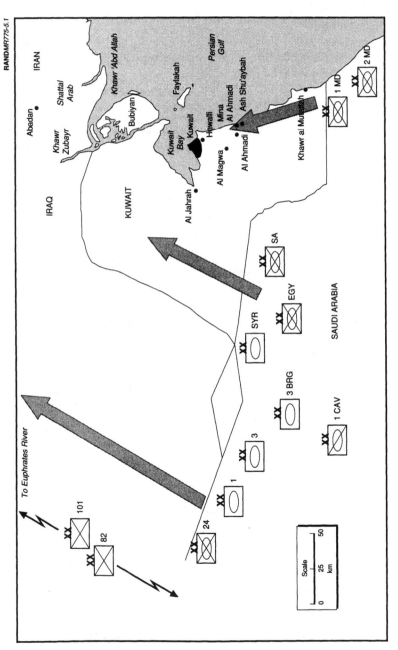

Figure 5.1—Schwarzkopf's Concept

DESERT STORM. The President approved the NSD, and Powell and the Secretary of Defense signed the execute order at 5 p.m. Washington time, authorizing General Schwarzkopf to initiate Operation DESERT STORM at 3 a.m. Riyadh time on January 17.[20]

## THE CAMPAIGN

On January 17 at 1:30 a.m. in the Persian Gulf, the USS *Bunker Hill* fired a Tomahawk missile, the first of 106 Tomahawks that would be launched into Iraq during the first 24 hours of the war. The air campaign moved into high gear, achieving air superiority, blinding Iraqi C2 systems, and attacking strategic targets. Schwarzkopf decided to transition to Phase 3 on the fifteenth day of the bombing campaign:

> After two weeks of war, my instincts and experience told me that we'd bombed most of our strategic targets enough to accomplish our campaign objectives; it was now time, I thought, to shift most of our air power on to the army we were about to face in battle.[21]

By February 8, the two corps had almost completed their move to the west and were occupying attack positions. Schwarzkopf calculated that the mountains of logistics material and the remaining units would be in place within ten days. He informed Secretary Cheney, who was in Riyadh for a briefing, that he would be ready to go anytime after February 21. On February 24 at 4 a.m., the coalition forces attacked on the ground.

Operation DESERT STORM lasted 42 days. The three air phases took 38 days. The Iraqi air defenses, command and control centers, and air forces were quickly neutralized. Many strategic targets, including some in Baghdad itself, were successfully attacked. An improvised but very effective attack against a deep-underground Iraqi C2 bunker produced an unexpected and embarrassing number of civilian casualties among those who had supposedly taken refuge there. The Iraqis employed ballistic missiles in attacks upon Saudi Arabia and Israel. Although those attacks proved of little military significance, their political consequences proved very distracting to the air effort.

---

[20]Gordon and Trainor, 1995, p. 206.

[21]Schwarzkopf and Petre, 1992, p. 430.

The coalition efforts to defend against those ballistic-missile attacks were mirror images—of more political than military effect. The bombing of Iraqi forces in Kuwait was relentless, but was probably more demoralizing that lethal in its effects.

The ground war lasted just 100 hours before President Bush, in consultation with his military commanders, called a halt. Of 42 Iraqi divisions in the theater at the beginning of the war, 27 were destroyed and an additional six were rendered combat-ineffective.[22]   However, over half the Republican Guard, including the nearly intact Hammurabi Division, escaped the enveloping "left hook," leaving a legacy of controversy about whether General Frederick Franks' VII Corps had moved quickly enough.[23]   When the Iraqis finally fled from Kuwait, they jammed the road north out of the city with vehicles and booty, which the coalition air power then blocked and savaged. The vivid descriptions of the resulting carnage were probably a significant factor in the decision to halt military operations.

## COMMAND AND CONTROL

From the perspective of the U.S. military, the chain of command finally seemed to work as it was supposed to—but too often previously had not. In Schwarzkopf's words,

> the President had been presidential; the Secretary of Defense had concentrated on setting military policy; the Chairman of the Joint Chiefs had served as the facilitator between civilian and military leadership; and as theater commander I'd been given full authority to carry out my mission.[24]

---

[22]Schwarzkopf and Petre, 1992, p. 467.

[23]Criticism of General Franks persists, although an Army study conducted after the war showed that VII Corps was only 10 hours behind schedule in engaging the Republican Guard, a reasonable performance considering the technical difficulties of organizing the logistics support of such a movement.  Further, XVIII Corps (McCaffrey's 24th Mechanized and Peay's 101st Airmobile Divisions) was positioned to interdict the bulk of the fleeing forces, but was forestalled from doing so by the higher-level wrangling over when to effect a cease-fire. See Gordon and Trainor, 1995, pp. 405–409, 429.

[24]Schwarzkopf and Petre, 1992, p. 368.

One result of this fidelity to the command structure was that, after launching the ground offensive, Powell had to wait until Schwarzkopf had the time to inform him of progress on the battlefield. At the same time, Schwarzkopf understood that *he* was not going to be able to track the entire battle in real time—only key portions of it:

> Back at the war room in Riyadh, we were so removed from the action that all we knew was that our forces were finally on their way across the border. It might take an entire day to piece together an accurate picture of how the attack was progressing. I desperately wanted to do something, *anything*, other than wait, yet the best thing I could do was stay out of the way. If I pestered my generals, I'd just distract them: I knew as well as anyone that commanders on the battlefield have more important things to worry about than keeping higher headquarters informed. . . . My job was to stay in the basement with our radios and telephones, assessing the offensive as it developed, keeping the senior commanders apprised of one another's progress, and making sure we accomplished three strategic goals: to kick Iraq out of Kuwait, to support our Arab allies in the liberation of Kuwait City, and to destroy the invading forces so Saddam could never use them again.[25]

Nonetheless, Schwarzkopf listened attentively to the electronic "sounds" of the battlefield as events developed. At about noontime, eight hours after the initiation of the ground campaign, he received news that the Iraqis had destroyed the desalinization plant in Kuwait City by blowing it up:

> Since Kuwait City had no other source of drinking water, this could only mean that the Iraqis were about to leave. And if they intended to pull out of Kuwait City, I reasoned, they intended to pull out of Kuwait.
>
> At that point, I knew that I had to act. Timing is everything in battle, and unless we adjusted the plan, we stood to lose the momentum of the initial gains. *I'd fought this campaign a thousand times in my mind,* visualizing all the ways it might unfold, and from the fragmentary reports coming into the war room, I could discern that

---

[25]Schwarzkopf and Petre, 1992, p. 452.

the Iraqis were reeling.  If we moved fast, we could force them to fight at a huge disadvantage.[26]

This was the only significant intervention that Schwarzkopf made during the course of the ground campaign.  As a result of the evidence of the Iraqi withdrawal, he sprang the main attack (the "left hook") approximately 18 hours early.

## SCHWARZKOPF'S COMMAND CONCEPT

Although the strategic war aims of the United States were never explicitly spelled out by President Bush, General Powell instructed Schwarzkopf, in early December, to draft a Strategic Directive, which is reproduced below.

### DRAFT PROPOSED STRATEGIC DIRECTIVE TO COMBINED COMMANDER

1. TASK.  Undertake operations to seek the complete withdrawal of Iraqi forces from Kuwait in accordance with the terms of the UN resolutions and sanctions.  If necessary and when directed, conduct military operations to destroy Iraqi armed forces, liberate and secure Kuwait to permit the restoration of its legitimate government, and make every reasonable effort to repatriate foreign nationals held against their will in Iraq and Kuwait.  Promote the security and stability of the Arabian/Persian Gulf region.

2. AUTHORIZATION.  When directed, you are authorized to conduct air operations throughout Iraq and land and sea operations into Iraqi territory and waters as necessary to liberate and secure Kuwait and destroy Iraqi forces threatening the territory of Kuwait and other coalition states.  Forces should be prepared to initiate offensive operations no later than February 1991.

At any time, you are authorized to take advantage of full or partial withdrawal of Iraqi forces from Kuwait by introducing forces under your command to secure Kuwaiti territory and waters, defend

---

[26]Schwarzkopf and Petre, 1992, p. 453 (emphasis added).

against renewed aggression, and permit the restoration of the legitimate government in Kuwait.

Pending authority to execute operations to destroy Iraqi forces and liberate Kuwait, defend Saudi Arabia. Should Iraqi forces invade Saudi Arabia, you are authorized to conduct air, land, and sea operations throughout Kuwait and Iraq, their airspace, and territorial waters.

3. OPERATIONAL GUIDANCE. The objectives of your offensive campaign will be to destroy Iraqi nuclear, biological, and chemical production facilities and weapons of mass destruction; occupy southeast Iraq until combined strategic objectives are met; destroy or neutralize the Republican Guard Forces Command; destroy, neutralize, or disconnect the Iraqi national command authority; safeguard, to the extent practicable, foreign nationals being detained in Iraq and Kuwait; and degrade or disrupt Iraqi strategic air defenses.[27]

Schwarzkopf's command concept is clearly derived from and serves the strategic objectives enumerated earlier. Recast in our format of an *ideal* command concept, it might read as follows:

## I. ABOUT THE ENEMY AND HIS PLANS:

1. The enemy [Iraq] currently has approximately 400,000 troops in the Kuwaiti Theater of Operations (KTO). He expects us [the United States] to conduct amphibious and ground operations aimed at the recapture of Kuwait City and the liberation of Kuwait.

2. The enemy is expected to resist a frontal attack fiercely. Once flanked and isolated, however, resistance in the KTO should quickly collapse.

3. You should expect the Iraqis to attempt to inflict as many casualties as possible on our forces, possibly through the use of chemical or biological agents.

---

[27]Schwarzkopf and Petrie, 1992, pp. 386–387.

4. The Iraqis will likely respond to tactical surprise by attempting to preserve the "center of gravity" they have vested in the Republican Guard divisions.

## II. ABOUT OUR FORCE DISPOSITIONS AND PLANS:

1. We shall first attack with air power to incapacitate Iraqi command, control, logistics, and air defense systems. We shall follow this with an intensive air campaign to keep in place, disrupt, attrit, and demoralize deployed Iraqi forces in the KTO. We shall then attack with four army corps to (1) neutralize Iraq's fielded forces, (2) liberate the country of Kuwait, and (3) destroy Iraq's ability to conduct invasion operations in the future.

2. We shall conduct this operation in four phases, the first three with air power and the last with combined forces:

Phase 1—Using strategic and tactical air assets, achieve air supremacy in the KTO and incapacitate Iraq's command and control system.

Phase 2—Extend the air war to destroy, disrupt, and render ineffective Iraq's warmaking ability, placing top priority on destroying weapons of mass destruction (nuclear, biological, and chemical), the eight Republican Guard divisions, and petrochemical facilities.

Phase 3—Having isolated the theater, conduct an intensive bombing campaign against fielded Iraqi forces in Kuwait proper, with the aim of disrupting, demoralizing, and destroying as many of the 400,000 troops occupying Kuwait as possible.

Phase 4—Having attrited, disrupted, and demoralized the Iraqi Army, conduct a rapid and violent ground campaign designed to surround, isolate, and defeat completely the Iraqi occupation forces in Kuwait.

## III. ABOUT CONTINGENCIES:

1. If the Iraqi forces give evidence of withdrawal from the theater at any time—during the air campaign or during the subsequent

ground campaign—we shall accelerate our planned operations, with the aim of destroying their forces during the confusion of withdrawal.

2. If the air campaign is delayed by weather or other impediments, we shall adjust our ground campaign accordingly.  We shall delay our repositioning to the west until we are assured that the Iraqis have been blinded and that any countermove by their forces can be exploited by air strikes against their forces on the move.

## ASSESSMENT

This modern-day blitzkrieg offers an insight from a vantage point different from Guderian's.  Schwarzkopf did not have (nor did he seek) "Guderian's perspective"—at least not physically.  *Conceptually*, however, Schwarzkopf clearly understood and identified the information that was essential to managing the execution of his command concept.  Schwarzkopf sitting in his bunker and reacting to the destruction of the water-desalinization plant is perhaps one of the clearest examples of the theory that history offers.  Unlike French Field Marshal Joseph-Jacques-Césaire Joffre in his chateau, Schwarzkopf, although physically isolated, was mentally tuned in to *the way the battle had to go*.  Powerful evidence of this is the minimal level of traffic over command channels between Schwarzkopf and his field commanders during the battle.  Schwarzkopf essentially listened in on the command networks, mentally ticking off the progress of the battle against his own expectations, intervening when he (correctly) detected activity at variance with his expected timetable.  Indeed, except for the decision to advance the timing of the "left hook," Schwarzkopf could have left the theater to his subordinates to carry out his plans.

In one sense, Schwarzkopf can be criticized for not executing his command concept with complete success.  Many have commented on Schwarzkopf's handling of war termination in this conflict, accusing the general of having lost touch with the status of the Republican Guard divisions and recommending ending the conflict before the retreating Guard divisions were enveloped and rendered combat-ineffective.[28] In truth, it is possible that the unexpected rapidity with which the Marines advanced on the right accelerated events beyond Schwarzkopf's ability to precisely control them. The failure to completely destroy the Republican Guard is probably the result of this and two other factors: (1) the unexpectedly light resistance, low casualties, and obvious destruction of Iraqi forces in place, which undoubtedly made a precise calculation of when to terminate a difficult one and show Schwarzkopf's expectations to be too pessimistic, and (2) the relatively vague political objectives set by the U.S. leaders, which never really specified how far they wanted Schwarzkopf to go beyond the liberation of Kuwait. Schwarzkopf *assumed* that destruction of the Republican Guard would be necessary to the liberation of Kuwait—and when it was clear that it was not, other considerations intervened (concern about Arab reaction to the wanton slaughter of Iraqi forces, for instance) to force what may have been, in retrospect, a premature termination of the conflict.

In that light, it is important to separate Schwarzkopf's generalship from his articulation and management of his command concept. His failure to coordinate the planning and execution of the Marine and Army operations, noted in several of the sources, probably allowed the Marines to push the Republican Guard out of the trap before it closed. That said, Schwarzkopf's accomplishment is nonetheless impressive. He developed a vision, communicated it effectively to his subordinates, and employed his C2 resources to give him the information he believed was necessary to make critical decisions during the war. We can also say that his C2 system fully supported his command concept—a support that might have been more obvious if the initial attacks had met with difficulty. He understood how to use his capability, and focused his ability to look "everywhere" on looking at areas that were essential to the confirmation or refutation of his plan. Given the fact that Schwarzkopf's *need* to communicate was minimal because his plan was basically sound, a more difficult enemy would not have significantly altered Schwarzkopf's ability to listen for key events and understand when it was time to make a decision.

---

[28]Gordon and Trainor (1995) are especially critical of this.

Unlike Schwarzkopf, MacArthur did not listen for the key event; he built his plan around what he knew beforehand to be a key factor: the absolute necessity of landing at Inchon. Like Guderian, he knew that a hard drive was essential to vanquishing the enemy; unlike Guderian, he had only one month in which to practice, not five or six.

# Chapter Six

# THE VISIONARY: MACARTHUR AT INCHON

> Rapidity is the essence of war; take advantage of your enemy's unreadiness, make your way by unexpected routes, and attack unguarded spots.
>
> —Sun-Tzu

## BACKGROUND

Following a series of reverses in the early days of the Korean War, the U.S. Eighth Army won its first major victory against the North Korean forces by executing a brilliant amphibious landing at Inchon. Leapfrogging up the western coast of Korea and striking deep into the rear of the North Korean Army (NKA), this operation unhinged the North Koreans' advance, caused their offensive campaign to collapse, and forced their army into headlong retreat. If not for the later intervention of the Chinese, the Inchon landing would almost certainly have resulted in the decisive defeat and collapse of the North Korean government, effectively ending the conflict.

The architect of this victory was General Douglas MacArthur, the 70-year-old veteran of both World Wars, who at the time was the Commander in Chief, Far East (CINCFE) Theater. In this capacity, MacArthur was responsible for crafting a response to the North Korean invasion, which had begun on June 25, 1950. While MacArthur was throwing forces into the fight as quickly as possible in an attempt to stop the North Korean onslaught, his mind was clearly set on finding a decisive solution to the conflict:

> During the first week of July, with the Korean War little more than a week old, General MacArthur told his chief of staff, General [Edward M.] Almond, to begin considering plans for an amphibious

operation designed to strike the enemy center of communications at Seoul, and to study the location for a landing to accomplish this.[1]

By late July, the plan had taken shape:

> On 23 July, General [Edwin K.] Wright [MacArthur's G-3] upon MacArthur's instructions circulated to the GHQ [General Headquarters] staff sections the outline of Operation CHROMITE. CHROMITE called for an amphibious operation in September and postulated three plans: (1) Plan 100-B, landing at Inchon on the west coast; (2) Plan 100-C, landing at Kunsan on the west coast; (3) Plan 100-D, landing near Chumunjin-up on the east coast. Plan 100-B, calling for a landing at Inchon with a simultaneous attack by the Eighth Army, was favored.[2]

On August 12, MacArthur issued CINCFE Operation Plan 100-B and specifically named the Inchon-Seoul area as the target that a special invasion force would seize by amphibious assault. MacArthur began planning and preparation with a planning cell located in his own GHQ staff; forces earmarked for the operation were to be placed in GHQ Reserve as they became available.

On August 26, MacArthur requested and received authorization to activate a Corps Headquarters, X Corps, as the operational unit to execute the mission. On the same day, he appointed his Chief of Staff, Major General Almond, to command the corps.[3] The major ground units of X Corps were to be the 1st Marine Division, commanded by Major General Oliver P. Smith, and the 7th Infantry Division, commanded by Major General David G. Barr.

The 1st Marines were in the United States, and had to be swiftly outfitted and transported to the theater of operations. General Smith was given three weeks to get his division (less one regiment) ready, and at the last moment was given an additional replacement regi-

---

[1] Roy E. Appleman, *South to the Naktong, North to the Yalu*, Washington, D.C.: Center for Military History, 1961, p. 488.

[2] Appleman, 1961, p. 489.

[3] Appleman, 1961, p. 490.

ment composed of recruits and a battalion of Marines extracted from duty in the Mediterranean.[4]

The 7th Infantry Division, stationed in Japan at the time, was also undermanned, being at about half-strength. Its ranks had been thinned over the preceding two months to provide trained replacements for the 24th and 2nd Infantry Divisions in Korea.[5] As a stop-gap measure, over 8,000 Korean recruits were sent to Japan to augment the 7th Infantry Division.[6] Integrating, acclimating, and training these replacements was begun in a desperate hurry.

While MacArthur never wavered in his commitment to the Inchon plan, circumstances forced a series of postponements. Since the beginning of the war on June 25, the North Korean Army had forced the combined Republic of Korea (ROK)/U.S. forces to withdraw nearly 150 miles south—from positions near the Han River to defensive positions on the southern tip of the Korean peninsula, known collectively as the "Pusan Perimeter" (see Figure 6.1). Stopping this advance had absorbed every soldier, sailor, and airman that MacArthur could get his hands on.

During the waning days of August 1950, the North Korean High Command also had problems—exploiting this success was proving difficult. With most of its forces deep in South Korea, North Korea's primitive supply system was overstrained. Especially troublesome was the inability of the North Korean Army to feed and clothe its soldiers. By September 1, many units were having trouble feeding their troops more than one meal per day.[7] With winter not far away, time was growing short for the North Koreans to finish the fight.

In response to this urgency, the North Korean High Command planned a massive offensive for early September aimed at placing pressure on all sides of the Pusan Perimeter. For this effort, the

---

[4]Appleman, 1961, p. 491.

[5]On July 27, the 7th Infantry Division was 9,117 men under strength (Appleman, 1961, p. 491).

[6]Appleman, 1961, p. 492.

[7]Appleman, 1961, pp. 393–394.

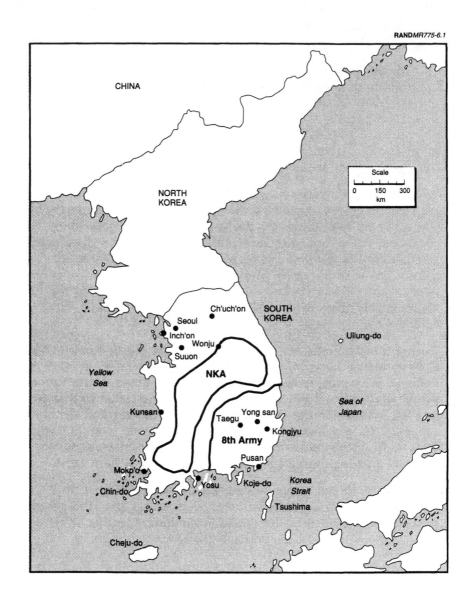

**Figure 6.1—The Pusan Perimeter**

North Koreans assembled 13 infantry divisions and 2 armored divisions, with a total strength of nearly 98,000 men.[8]

This force attacked on September 1, immediately placing the integrity of the Pusan Perimeter in peril. By September 3, General Walton H. Walker, the Eighth Army commander, had crises in at least five places around the perimeter.[9] After several desperate battles at Kongjyu in the east, Taegu in the center, and Yongsan in the north, the combined Eighth Army/ROK forces managed to hold, and by September 11, those crises had passed.[10]

Two weeks of the heaviest fighting of the war had produced inconclusive results. However, with most of the enemy's combat power piled up near Pusan, MacArthur clearly saw that his opportunity had come, despite the fact that the Eighth Army remained in dire peril. The frustrating wait was over.

Before he could launch his invasion, however, MacArthur had other battles to fight. The enemy was not the only obstacle to the execution of his plan. The selection of the landing site was a subject of great controversy within the U.S. military. MacArthur had a clear vision of the importance of Inchon as a landing site, but the practicalities of battling the difficult tidal conditions there caused the Navy and Marine Corps to raise strong objections. The issue had come to a head at a conference in Japan on July 23 attended by General J. Lawton Collins, the Army Chief of Staff, Admiral Forrest P. Sherman, the Chief of Naval Operations, and General MacArthur:

> He [MacArthur] talked as though delivering a soliloquy for forty-five minutes, dwelling in a conversational tone on the reasons why the landing should be made at Inchon. He said that the enemy had neglected his rear and was dangling on a thin logistical rope that could be quickly cut in the Seoul area, that the enemy had committed practically all of his forces against the Eighth Army in the south and had no trained reserves and little power of recuperation.[11]

---

[8]Appleman, 1961, pp. 394–395.

[9]Appleman, 1961, p. 397.

[10]Appleman, 1961, p. 487.

[11]Appleman, 1961, p. 493.

In his arguments, MacArthur focused on the reasons for making such a landing and the conditions under which a landing would achieve the objectives of cutting the supply lines of the NKA, destroying the NKA (see Figure 6.2):

> MacArthur stressed the strategic, political, and psychological reasons for the landing at Inchon and the quick capture of Seoul, the capital of South Korea. . . . General MacArthur then turned to a consideration of a landing at Kunsan, 100 miles below Inchon, which General Collins and Admiral Sherman favored. MacArthur said the idea was good but the location wrong. He did not think a landing there would result in severing the North Korean supply lines and destroying the North Korean Army. He returned to his emphasis in Inchon, saying that the amphibious landing was tactically the most powerful device available to the United Nations Command and that to employ it properly meant to strike deep and hard into enemy territory.[12]

Collins and Sherman returned to Washington with the understanding that the landing site was still an open question. MacArthur, however, was unswerving in his preparations for Inchon alone. For the next month, despite instructions from Washington to prepare plans for an alternative site as well as for Inchon, MacArthur proceeded with his original intention and issued his United Nations Command operations order for the landing on August 30.

It was at this time that the North Korean Army initiated its massive assault on the Pusan Perimeter. The Eighth Army commander voiced doubts about whether the perimeter could hold, to say nothing of conducting an offensive breakout in support of the landings to the north. MacArthur insisted that the opportunity was there, that it would be strategically foolish to reinforce the Pusan Perimeter, and on September 6 issued an instruction to all his major commanders setting the landing date for September 15. Because of his insistence and his powers of persuasion with the Joint Chiefs, MacArthur received authorization on September 9 to proceed with the landing at Inchon—despite the fact that the encircled UN forces at Pusan were hanging on by a slender thread.

---

[12]Appleman, 1961, p. 493.

RAND*MR775-6.2*

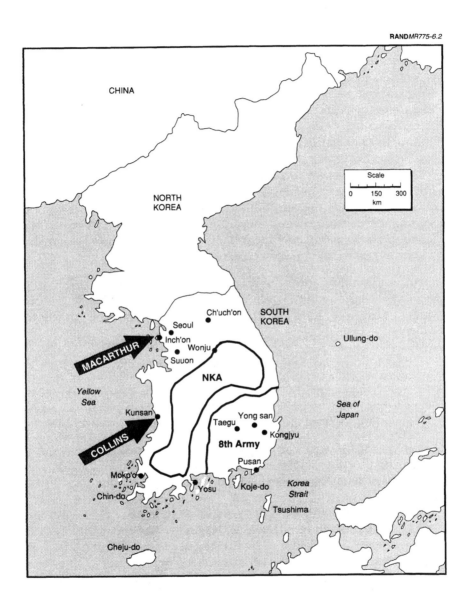

**Figure 6.2—Competing Command Concepts for the Relief of the
Pusan Perimeter**

## THE PLANS

It is revealing of MacArthur's vision that he considered the Inchon operation to be an *Army* mission. Although the Navy and Marine Corps were essential to the success of the operation, MacArthur understood that the real task—the destruction of the North Korean Army—would have to be accomplished by the Eighth Army's "hammer" slamming the NKA against the X Corps' "anvil" 30 km inland from the beach. Additionally, intelligence estimates predicted little or no resistance to the landing operation, and only small numbers of NKA forces available to oppose the recapture of Seoul. For this reason, while the amphibious operation required an enormous amount of preparation and planning under a severe time constraint, and a high degree of technical competence to execute, MacArthur seemed to consider it a low-risk exercise.

Despite grumbling from the Navy and the Marines, MacArthur activated an Army headquarters and placed his own deputy in command. The Army focus of the operation was evident from the top down:

> [MacArthur] would require a Headquarters for a lightning-like strike which could be handled by the personnel then available to the Far East Command [with] many able, experienced, and senior officers who had commanded in Europe and in Italy. *Once a force had been landed* the principal problem would be a land operation over some 18 miles of terrain involving a river crossing. . . . *The real essence of the Inchon landing was not merely to land and form a beachhead* but to drive across difficult terrain 18 miles and capture a large city and thereafter properly outpost and protect that city.[13]

The complexity of the landing operation is reflected in the command relationships shown in Figure 6.3. MacArthur placed Admiral Arthur D. Struble (Commander, 7th Fleet) in overall command of the amphibious operation. Struble task-organized his assets. Each of the resulting six task forces (TFs) has its own function. It is worth noting that the sole function of five of the task forces was to work together to get the sixth, TF 92/X Corps, ashore and on its feet. For

---

[13]Robert D. Heinl, *Victory at High Tide*, Philadelphia: J. B. Lippincott, 1968, p. 54 (emphasis added).

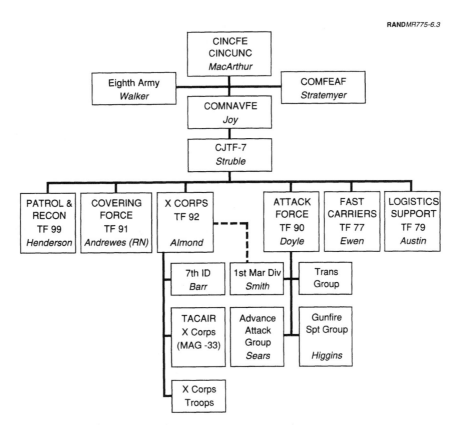

NOTE: CINCFE = Commander in Chief of the Far East Command; CINCUNC =
Commander in Chief of the United Nations Command; CJTF-7 = Commander, Joint
Task Force, 7th Fleet; COMFEAF = Commander of the Far East Command Air
Force; COMNAVFE = Commander of the Navy of the Far East Command; ID =
Intelligence Division; MAG = Marine Air Group; TACAIR = Tactical Air.

**Figure 6.3—Operation CHROMITE Command Relationships**

this purpose, the 1st Marine Division was detailed from X Corps to
Admiral James Doyle's attack force (TF 90) until such time as there
were sufficient forces ashore to activate X Corps command functions.

The landings themselves had to be made in the face of extraordinary
conditions in the Inchon harbor, where tides of 31 feet were not un-
common. Hydrographic studies showed that, on invasion day, there
would be a short morning and afternoon window at high tide during

which there would be enough water in the harbor to support the LSTs (Landing Ships, Tank), critical to rapid projection of combat power inland. Therefore the decision was made to conduct the landing in two phases. Moreover, the landing had to occur on September 15:

> The extreme high tide on that day—the only such day in September—would put the maximum high tide over Inchon's mud flats at 31.2 feet. Twelve days later, on the 27th, there would be only 27 feet (two feet short of what the LSTs needed). Not until October 11 would there again be 30 feet of water.[14]

The first phase was to be an early-morning assault by a Marine Battalion Landing Team (BLT) to secure the island of Wolmi-do, which guarded the approaches to the harbor. Then there would be a hiatus of several hours before the tides would permit the landing of the main force.

## THE BATTLE

On September 10 and 11, the flotilla departed for Inchon, the 1st Marines sailing from Kobe and the 7th Infantry Division from Yokohama, Japan. Despite the interference caused by two typhoons, the entire force managed to get under way on time.

On September 13 and 14, the Gunfire Support Group, commanded by Rear Admiral J. M. Higgins, conducted a preliminary bombardment in conjunction with aircraft from Task Force 77 to knock out the shore defenses around the landing sites. Having succeeded in their task, they retired to support positions.

The X Corps landing force arrived off Inchon on September 15, with a total strength of over 70,000 combatants. The assault force debarked into their landing craft, crossed the line of departure at 0625, and headed for the landing beaches on the first objective, the harbor island of Wolmi-do, approximately 1 mile away:[15]

---

[14]Heinl, 1968, p. 33.

[15]Appleman, 1961, p. 505.

The first troops ashore moved rapidly inland against almost no resistance. Within a few minutes the second wave landed. Then
came the LSV's carrying the tanks . . . the reduction of the island
proceeded systematically and was secured at 0750.[16]

The landing team on Wolmi-do settled in and prepared to support
the main landing across the harbor. At 1530, the tide was sufficiently
high to begin launching the landing craft for the main assault. The
first wave, led by the 5th Marine Regiment, breasted the seawall in
Inchon harbor at 1730. Despite several sharp encounters with North
Korean defenders, the landing force secured its D-Day objectives by
0130, September 16.[17]

Early on the morning of September 16, the two Marine regiments
ashore effected a linkup, sealing off the approaches to Inchon and
enabling the landing of a regiment of ROK Marines, who conducted
mop-up operations in the city. By the evening of September 16, the
situation had sufficiently developed for General Smith, the 1st
Marine Division Commander, to move his headquarters ashore and
notify the Assault Force Commander, Admiral Doyle, that he was
assuming responsibility for operations ashore.[18]

By September 17, the Marines had moved inland and captured
Kimpo Airfield, approximately 8 miles west of Seoul. On September
19, Marine aircraft were conducting flight operations from Kimpo,
completing the landing objectives, and setting the stage for the capture of Seoul. At the same time, the 7th Infantry Division had landed
and had assumed blocking positions to the south of the Seoul highway.

Despite difficult tidal conditions and the small size of the harbor, by
the evening of September 18, the Navy had unloaded 25,606 persons,
4,547 vehicles, and 14,166 tons of cargo. MacArthur's concept was
proven valid, the landing achieved strategic surprise, and the stage

---

[16]Appleman, 1961, p. 506.

[17]Appleman, 1961, p. 507.

[18]Appleman, 1961, p. 509.

was set for the liberation of Seoul and the destruction of the North Korean Army.[19]

## COMMAND AND CONTROL

OPLAN 100-B envisioned a command structure similar to that used by MacArthur during the Pacific Campaign of World War II, wherein screening, covering, fire support, and landing operations were to be conducted under the command of naval officers. Once a sufficient beachhead was established, the Attack Force Commander, Admiral Doyle, would relinquish ground command to General Smith, 1st Marine Division Commander. As soon as 7th Infantry Division began coming ashore, General Almond would activate his headquarters ashore and X Corps would direct all ground operations from that point forward, with Marine air and naval gun fire in direct support, and naval air in general support.

Although Admiral Struble was in overall command of the operation, the vision was clearly MacArthur's. He had transmitted this vision so compellingly and clearly that his contribution to directing the battle—the execution of his vision—consisted of remaining visible and offering words of encouragement. Observing the successful— although confusing and somewhat disorganized—landings at Red and Blue Beaches, MacArthur

> turning to Admiral Doyle, . . . directed, "Say to the Fleet [in an echo of Nelson?], 'The Navy and Marines have never shone more brightly than this morning,'" The admiral's pencil hovered until he knew from context whether it would be "shown" or "shone," then he finished the sentence and handed it to a staff officer. With a broad smile MacArthur glanced around at Generals Shepherd, Smith, and Almond, and the admiral, and said: "That's it. Let's go get a cup of coffee."[20]

---

[19]Appleman, 1961, p. 513.

[20]Heinl, 1968, p. 93.

## MACARTHUR'S COMMAND CONCEPT

In view of the assumptions embedded in OPLAN 100-B—that strategic surprise would be achieved, that air and naval supremacy would compensate for unforeseen difficulties, and that the real fight would commence once the landing force had moved inland for the liberation of Seoul—this plan could be used as a template for an ideal command concept. Such a concept, if written explicitly, might look as follows:

### I. ABOUT THE ENEMY AND HIS PLANS:

1. The enemy [the NKA] currently has no more than 6,500 troops in the Inchon-Seoul area. He does not suspect our [U.S. troops'] intentions.

2. The enemy is expected to press his advantage at Pusan with the intent of achieving a decisive victory before the winter snows. Essentially all combat-effective enemy troops are 150 miles south of Seoul, oriented toward the Pusan Perimeter.

3. You should expect the North Koreans to attempt to reinforce and/or assist the Seoul garrison with at most 3 divisions: the 3rd, 13th, and 10th. They will be unable to influence the situation in Seoul for at least two weeks after our landing. Our severing their supply line will undoubtedly degrade this capability further.

### II. ABOUT OUR FORCE DISPOSITIONS AND PLANS:

1. We shall isolate the Inchon/Seoul area with air and naval forces from Task Force 91 (the covering force), Task Force 99 (the patrol and reconnaissance force), and Task Force 77 (the fast carrier force) with the intent of preventing North Korean or other interference in landing operations.

2. We shall land three Marine regiments, followed by an Infantry division and three ROK regiments to (1) secure a beachhead, (2) capture Kimpo Airfield, (3) liberate Seoul, and (4) establish blocking positions astride the North Korean main line of communications, the Seoul-Taejon-Taegu highway. You have two weeks to accomplish this task.

3. Eighth Army will conduct a breakout and pursuit of the North Korean forces surrounding Pusan, forcing them to fall back on X Corps' blocking position. Given the severing of their main supply route, the North Korean Forces will have no choice but to fall back.

4. Our strategic objective is the relief of South Korea, but our tactical objective is to inflict as much damage as possible upon the withdrawing enemy army—commensurate with preserving the ability of X Corps to continue the fight. The measure of tactical success will be whether the North Korean forces are rendered incapable of counterattacking once driven from the country.

5. If X Corps is a rock upon which the withdrawing North Korean forces can be shattered, the enemy cannot prevail in his objectives. Therefore, priority of naval and air-delivered fires will be given to X Corps units in contact with retreating North Korean units, with the objective of destroying the North Korean units.

## III. ABOUT CONTINGENCIES:

1. Although Inchon is a technically difficult landing site, the opposition should be light, and we have adequate fire support to cover our attempts if they should be complicated by difficulties with tides and obstacles.

2. Although resistance should be light, we must move quickly to capture Seoul and block the retreat of the North Korean forces in the south. The time is adequate, but enemy resistance and the risks of casualties should not be permitted to delay our movement to blocking positions.

3. Our naval and air forces must be prepared to support the landings at Inchon, to exploit our movement to capture Seoul, and to block the escape of North Korean forces in the south. Thereafter, they must be prepared to assist in the destruction of those North Korean forces.

4. Despite its desperate position within the Pusan Perimeter, the Eighth Army must be prepared to go from the defense to the offense as soon as the North Korean forces recognize their peril and begin to retreat.

5. Should the landing forces encounter enemy resistance heavy enough to disrupt the accomplishment of our strategic objective, the overriding imperatives are (1) to preserve the landing force and (2) to retain freedom of maneuver. The landing forces must be prepared to (1) fight their way inland to seize and hold defensible terrain or (2) conduct a fighting withdrawal and reembarkation supported by our naval and air forces.

## ASSESSMENT

MacArthur's insistence on the Inchon site was the key to the decisiveness of the victory. While the naval leadership and the Joint Chiefs supported the concept of "let's land where we can best land," MacArthur insisted on "let's land where we *must* land to achieve our strategic objectives":

> General MacArthur's two earliest and most important decisions were to (1) go to Inchon and (2) do it as soon as he could. Both decisions were embodied in Far East Command Operation Plan 100-B [CHROMITE], which MacArthur issued on August 12, 1950.[21]

MacArthur's plan, despite the technical difficulties inherent in its landing operations, was strategically and operationally sound. Many assumptions underlay the plan, but most were made apparent and were argued out during the planning phases. The effectiveness of his advocacy of this particular plan showed that he had a clearly developed command concept, one that was, in turn, articulated to and internalized by subordinate leaders. In placing the Army at the center of his plan, MacArthur in his task organization, or structure, clearly demonstrated his understanding of the strategic significance of his concept, and clearly communicated his intent that this was more than just a "mere landing" or "diversionary attack." Overall, the effectiveness with which MacArthur developed and articulated his concept is shown by his singular contribution to command on D-Day: "That's it. Let's go get a cup of coffee."

MacArthur was successful largely because, although his command concept was not congruent with those of higher commands, he was

---

[21]Heinl, 1968, p. 33.

ultimately able to convince those commands of the wisdom of his assumptions. In the next chapter, we present a case study of a commander whose concept, in contrast, relied on the false and invalid assumptions made by higher commands.

# Chapter Seven

# NO TIME FOR REFLECTION: MOORE AT IA DRANG

Nothing except a battle lost can be half so melancholy as a
battle won.

—Wellington, Dispatches from Waterloo, 1815

## BACKGROUND

During November 14–16, 1965, the 1st Battalion, 7th Cavalry
Regiment, 1st Cavalry Division of the U.S. Army fought a battle in the
Central Highlands of Vietnam that fundamentally changed the
character of the war. Also a very bloody conflict—the bloodiest of the
war—the four days of fighting caused the 7th Cavalry Regiment to
suffer a higher percentage of casualties than had any regiment,
Union or Confederate, at Gettysburg.[1]

On February 6, 1965, an attack by Viet Cong forces on the U.S.
Advisor compound at Pleiku left eight Americans dead and over one
hundred wounded. This was a serious escalation of the war. Up
until that time, U.S. involvement had been characterized by small
counterinsurgency operations conducted by Special Forces and
Army advisers throughout the country. These operations were oriented primarily toward eliminating local support for the Viet Cong
and enabling local villagers to defend themselves.

In response, General William C. Westmoreland, Commander of U.S.
Forces in South Vietnam, asked for, and got, a battalion of Marines to
guard the airbase at Pleiku. In April, he asked for two more battalions of Marines and permission to transition from strictly defensive
duties to actively seeking to engage the Viet Cong. The Marines, plus

---

[1]Harold G. Moore and Joseph L. Galloway, *We Were Soldiers Once . . . and Young,* New
York: Random House, 1992, p. xx.

the Army's 173d Airborne Brigade and a newly designated helicopter-borne division, the 1st Cavalry Division (Airmobile), were mobilized and deployed to Vietnam in the summer of 1965.[2]  In late August, the 1st Cavalry Division moved into permanent quarters at An Khe in the geographic center of South Vietnam's Central Highlands (see Figure 7.1).

RAND*MR775-7.1*

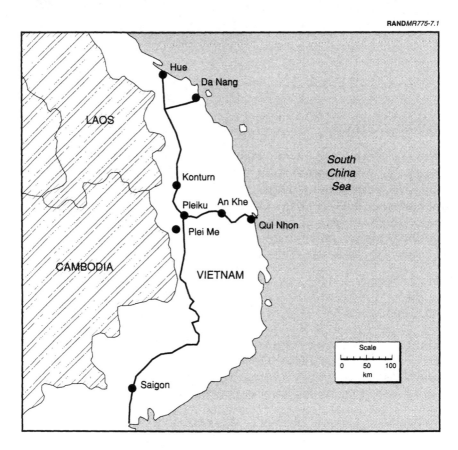

**Figure 7.1—II Corps' Tactical Zone**

---

[2]Moore and Galloway, 1992, pp. 14–15.

At the same time, North Vietnamese General Chu Huy Man, Commander of the Western Field Front Headquarters, ordered preparations for conducting a general offensive to wrest control of a major part of the Central Highlands (Kontum, Pleiku, Bin Dinh, and Phu Bon Provinces) from the South Vietnamese government. His objectives were to destroy Special Forces Camps at Plei Me, Dak Sut, and Duc Co and the district headquarters of the Saigon government at Le Thanh, and to capture the city of Pleiku. To accomplish these tasks, he had at his disposal three regular North Vietnamese Army (NVA) regiments, the 32d, 33d, and 66th.[3]

In mid-October, shortly after the 1st Cavalry Division moved into its new quarters at An Khe, General Man kicked off his campaign with an assault on the Special Forces camp at Plei Me. This attack, conducted by the 32d and 33d NVA regiments, was unsuccessful, largely as a result of the Americans' skillful use of fire support. General Man's regiments withdrew toward the Cambodian border, having stirred up a hornet's nest of American activity in response. For the next two weeks, troopers of the 1st Cavalry Division's 1st Brigade pursued and harassed the withdrawing NVA units in intensive search-and-destroy and reconnaissance operations.

As October drew to a close, it became clear to the division commander, Major General Harry W.O. Kinnard, that the NVA forces being pursued by the 1st Brigade were in danger of slipping across the Cambodian border, which was only 25 miles from Plei Me. He directed his attention toward the Chu Pong Massif, a rugged and remote piece of high ground straddling the Vietnam-Cambodian border—specifically, to the area between the foothills of the massif north to the Ia Drang River—and ordered his Third Brigade commander, Colonel Tim Brown, to search westward toward the Cambodian border (see Figure 7.2).

At this time, General Man was attempting to compensate for the heavy NVA losses at Plei Me by reorganizing for a renewed assault on the compound. The spot he picked for assembling and rehearsing his forces was the same piece of ground that Colonel Brown had

---

[3]John A. Cash, John Albright, and Allan W. Sandstrum, *Seven Firefights in Vietnam*, New York: Bantam, 1985, p. 5.

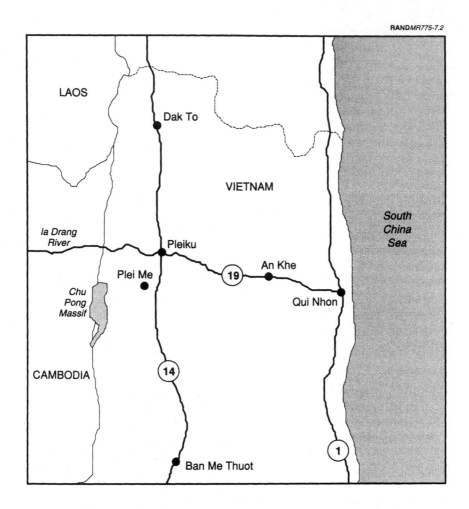

Figure 7.2—Ia Drang Area of Operations

been tasked to search. Thus, the stage was set for the unintended collision of two large opposing forces—the U.S. 1st Cavalry Division and the division-sized North Vietnamese B-3 Front.[4]

---

[4]Cash et al., 1985, p. 4.

## THE PLANS

General Man's operational concept was fairly simple. North Vietnamese strategic objectives would be well served by a war of attrition: By denying victory to the U.S./ARVN (Army of the Republic of Vietnam) forces, the North Vietnamese would eventually win. By raising the casualty rate, this process could be accelerated. Therefore, General Man's objective was to inflict as many casualties as possible on (1) the Special Forces garrison at Plei Me, (2) the ARVN relief column that would certainly be sent to their rescue, and (3) the U.S. force that would be sent to rescue the first two:

> "We wanted to lure the tiger out of the mountain," General Man said, adding: "We would attack the ARVN—but we would be ready to fight the Americans. . . . Headquarters decided we had to prepare very carefully to fight the Americans. Our problem was that we had never fought Americans before and we had no experience fighting them. We wanted to draw American units into contact for the purposes of learning how to fight them. We wanted any American combat troops; we didn't care which ones."[5]

The American forces were driven by a much different, reactive concept. They viewed themselves as exterminators called in to eradicate a particularly troublesome pest, rather than as military-political operators whose aim was to destroy the will of the North Vietnamese to fight. Therefore, when NVA forces were located, the immediate objective was to engage and destroy them. The implicit notion behind the American strategy was that, if this tactical objective could be achieved a sufficient number of times, then the overall strategic goal of maintaining a political order in Saigon friendly to U.S. interests could be achieved.

This unarticulated strategic concept served as the basis for the operational command concepts of the U.S. forces throughout the Vietnam War. The U.S. approach is well illustrated by the following passage from an account of a fight at Dak To in 1967:

---

[5]Moore and Galloway, 1992, p. 15.

> Regardless of the risks involved in attacking the enemy on terrain of his own choosing, the rare opportunity to catch the North Vietnamese in any concentration of forces could not be passed up.[6]

The assumption built into this concept was the idea—an arrogant one in retrospect—that finding engagement opportunities was the problem, not defeating an enemy under any terms of engagement. That assumption was remarkably persistent throughout the war, even in the face of mounting evidence to the contrary.

Thus, at 1700 hours on November 13, Colonel Brown ordered his 1st Battalion Commander (1/7 Cav), Lieutenant Colonel Harold G. Moore, to execute an airmobile assault into the Ia Drang Valley northeast of the Chu Pong Massif early on the morning of the 14th, and to conduct search-and-destroy operations through November 15. He would be allocated 16 helicopters for the assault. Fire support would be provided by two batteries (12 guns) from the 1st Battalion, 21st Artillery. They would be firing in support from Landing Zone (LZ) FALCON, 9 km east of the search area.[7] Helicopter gunships with rocket artillery and fixed-wing tactical air (Air Force) support would also be on call.[8]

As Moore began planning, he decided that, with a potentially large enemy force in the area, it would be safest to put his entire battalion into a single LZ, where he could concentrate his combat power most effectively. An aerial reconnaissance on the 14th confirmed that, of four potential sites, only one site, LZ X-RAY, would be suitable, since it could handle about eight ships at one time. This capacity would enable Moore to put about one rifle company on the ground with each flight, landing in two lifts each.

Having chosen X-RAY, Moore briefed his company commanders that intelligence estimates placed an enemy battalion approximately 5 km northwest of X-RAY, another force of undetermined size just south of X-RAY, and a hidden base approximately 3 km to the north-

---

[6]Allan W. Sandstrum, "Three Companies at Dak To," in Albright et al., *Seven Firefights in Vietnam*, Washington, D.C.: Office of the Chief of Military History (OCMH), 1970, p. 88.

[7]Cash et al., 1985, p. 5.

[8]Cash et al., 1985, p. 7.

west (see Figure 7.3). The 1/7 Cavalry would conduct an air assault into X-RAY, then search for and destroy enemy forces in the area, concentrating on streambeds and ridgelines.[9]  Once the battalion was on the ground, it would leapfrog forward by companies to the west, seeking contact with the enemy.

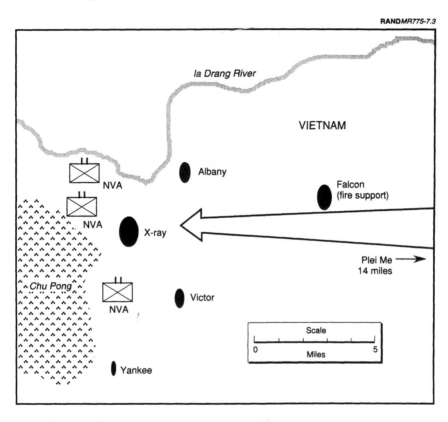

Figure 7.3—Moore's Plan

---

[9]Cash et al., 1985, p. 9.

## THE BATTLE

Following intense artillery preparation from his fire support base, LZ FALCON, Moore's B Company was airlifted into LZ X-RAY early on the morning of November 14. Sprinting from their helicopters, the soldiers of B Company swiftly secured the LZ without encountering any enemy. They settled in to await the 30-minute turnaround of the transport helicopters. As A Company was arriving around 1120, the aggressive probing of B Company turned up a prisoner, a North Vietnamese regular, who confirmed that there were three NVA battalions on Chu Pong Mountain, anxious to kill Americans but as yet unable to find them.[10]

Shortly thereafter, B Company found itself in a major contact: At least a company-sized NVA force was trying to overrun the LZ. It became difficult to land. Not until 1330 hours were A Company's last platoon and the lead elements of C Company able to land. The intensity of the NVA pressure on the LZ continued to increase. At 1420, the remainder of C Company and elements of D Company landed in the fifth lift. By 1500 hours, Moore estimated that he was opposed by 500–600 NVA, with more on the way.

Through skillful positioning of his forces and judicious use of reserves, Moore was able to beat back the initial onslaught and get the remaining tactical elements of his command into the fight. By 1630, D Company had landed and Moore threw them immediately into the fight to push back the attackers. However, General Man's forces persisted and continued to push against the American battalion. By 1700, Moore was fighting three separate actions: One force was defending X-RAY; two of his companies were attacking; and one platoon from Bravo Company was isolated, fighting for its life several hundred meters from the LZ (see Figure 7.4).[11]

During the night of the 14th, the 66th NVA Regiment moved its 8th Battalion south to the battle area and charged it with applying pressure against the eastern sector of X-RAY. Meanwhile, General Man

---

[10]Cash et al., 1985, p. 20.

[11]Cash et al., 1985, p. 24.

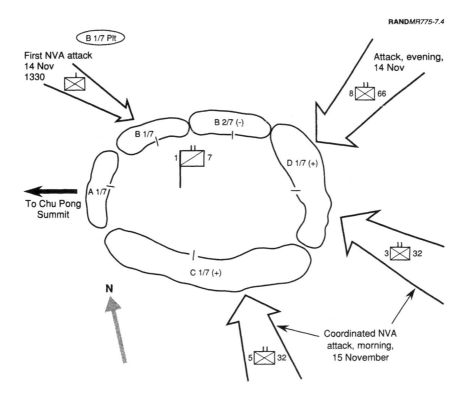

RAND*MR775-7.4*

B 1/7 Plt

First NVA attack
14 Nov
1330

Attack, evening,
14 Nov

8 ⊠ 66

B 1/7

B 2/7 (-)

1 ⊠ 7

D 1/7 (+)

To Chu Pong
Summit

A 1/7

3 ⊠ 32

C 1/7 (+)

N

5 ⊠ 32

Coordinated NVA
attack, morning,
15 November

**Figure 7.4—LZ X-RAY, November 14–15, 1965**

ordered the H-15 Main Force Viet Cong Battalion and the 32d NVA Regiment, some 12 km away, into the fight.[12]

During the morning of the 15th, C Company, holding the southern side of the X-RAY perimeter, came under a ferocious attack at first light. Shortly thereafter, D Company, holding the southeast quadrant of the perimeter, also came under heavy attack. The fighting was so close-in that enemy grazing fire was traversing the entire LZ. Using massive amounts of artillery and fire support, Moore's forces stalled and then repulsed the NVA attack. By 1000 hours, the NVA's

_____

[12]Cash et al., 1985, p. 26.

strong attempt to overrun the perimeter had failed and the attacks had ceased.[13]

Two hours later, three companies of 2/7 Cavalry arrived at X-RAY on foot to relieve Colonel Moore's battalion. Colonel Moore immediately directed an effort to relieve the lost platoon of Bravo Company, which had been cut off from the battalion for two days. This operation was successful. It recovered the surviving members of the platoon, as well as all of the American casualties.[14] The reinforced battalion settled in for its last night at X-RAY. By dawn of the 16th, after heavy probing throughout the night, the enemy attack had run its course.[15]

By 0930 on the morning of the 16th, the remainder of 2/7 Cavalry arrived at X-RAY, relieving the 1/7 Cavalry. Colonel Moore's unit marched overland to a nearby extraction zone and was airlifted back to An Khe.

## COMMAND AND CONTROL

General Kinnard's operational objective was to inflict losses on a fleeing enemy about whom hard information was scarce. Colonel Brown's own command concept reflected this objective, as well as the implicit assumption that Moore, properly supported, could handle whatever he encountered. Brown's intent, not very well expressed in his FRAGO (FRAGmentary Order), was roughly: "Find the enemy wherever he is and engage and destroy him. You have the force, training, and support to do the job."

During the airmobile insertion and afterwards, the C2 system was used overwhelmingly for control rather than command. Moore was decisively engaged from the moment he set foot on the LZ. All available bandwidth was devoted to killing those forces in front of him that were trying to destroy his battalion—that is, to alerting cavalry of where they were needed. Once Moore gained control of the situation, the C2 system, as he used it, was not adequate to support the

---

[13]Cash et al., 1985, p. 31.

[14]Cash et al., 1985, p. 36.

[15]Cash et al., 1985, p. 37.

development and articulation of a new, follow-on concept to destroy the remaining NVA forces.

## MOORE'S COMMAND CONCEPT

Although Moore did not explicitly articulate a command concept as such, he did give a very detailed ordcr. Like Admiral Jellicoe at the Battle of Jutland, he seems to have gone into battle more with implicit assumptions than with an explicit vision of what he should and could make happen. From the assumptions embedded in Colonel Brown's Brigade FRAGO—that tactical surprise would be achieved, that air and tube artillery support would compensate for unforeseen difficulties, and that the real fight would commence once 1/7 Cavalry had moved westward toward the Chu Pong Massif—Moore might have developed and communicated the following ideal command concept:.

## I. ABOUT THE ENEMY AND HIS PLANS:

1. The enemy [the NVA] currently has no more than 2,500 troops in the Chu Pong area. He does not suspect our intentions.

2. The enemy is expected to retire to the west, to sanctuaries in Cambodia, with the intent of reconstituting his strength for a renewed offensive in Pleiku and Bin Dinh Provinces.

3. You [U.S. troops] should expect the North Vietnamese to attempt to break contact and conduct a delay in-sector with two understrength regiments while continuing to withdraw most of his forces into Cambodia.

## II. ABOUT OUR FORCE DISPOSITIONS AND PLANS:

1. We shall conduct an air assault into LZ X-RAY and leapfrog up the Ia Drang Valley to isolate the NVA 32d Regiment as it continues its withdrawal toward the Cambodian border, with the objective of destroying it.

2. We shall land in seven lifts of two flights each over 2-1/2 hours.

3. Upon closure of the battalion into LZ X-RAY, 1/7 Cavalry will conduct a reconnaissance in the zone west to the military crest of the Chu Pong Massif, engaging and destroying enemy forces in-sector.

4. Our operational objective is the relief of enemy pressure in the Central Highlands, but our tactical objective is to destroy the 32d NVA Regiment—commensurate with preserving the ability of 1/7 Cavalry to continue to fight. The measure of tactical success will be rendering the 32d NVA Regiment combat-ineffective.

## III. ABOUT CONTINGENCIES:

1. Our air support and available artillery support will dominate chance encounters with any NVA force in the area of operations.

2. Our superior training, equipment, and support will enable us to fight and win against a larger NVA force.

3. We shall be deployed so that we can always assume a defensive posture that cannot be overrun by the NVA forces in the area, can be sustained by our superior fire support, and can be relieved by additional units from the 1st Cavalry Division.

This command concept was ideal only in the sense that, if it had been clearly articulated beforehand, Moore could have been absent and the outcome would likely have been about the same—except for the encouragement he lent his troops by his presence. But this concept was embedded in flawed higher-level concepts, concepts like those of the Royal Navy as it set forth to the Battle of Jutland, which also assumed that the salient problem was finding and engaging the enemy. If that could be accomplished, then the superior training, equipment, tactics, and doctrine would provide for success. At Ia Drang and Jutland, that arrogance resulted in costly stalemates.

## ASSESSMENT

The above command concept, mostly inferred from Brown's combat order, is basically tactical in nature—about how to move his forces into the fight, about how he was to be supported, and about what he hoped to accomplish by engaging the enemy—but very little about how the operation might unfold and what could be expected. These aspects seemed to have remained unspoken in training, both tactical and doctrinal. The order paints a clear enough picture of the objective of the mission and criteria for evaluating its success. But the tacit acceptance of the strategic assumptions and the tactical intentions of the enemy almost led to disaster at Ia Drang and ultimately led to a disaster in Vietnam.

The main flaw in the concept—such as it was—is that it did not really address the enemy's options. It assumed that the NVA had no capability to interfere with the accomplishment of Moore's objective. If the enemy is assumed not to be able to interfere, the only remaining problems are logistics- and terrain-related. In regarding the NVA as an essentially passive recipient of the violence Moore intended to inflict on it, Moore's concept did not allow for the contingency that ultimately arose. Moore's failure was *not* considering what his actions should be if the enemy did not meet his expectations. He neglected both to understand himself and to articulate to 1/7 Cavalry how to accomplish their objective in spite of the possibility of non-fulfillment of expectations. Those contingencies should have been a key component of an *ideal* command concept, and they are missing in Brown's FRAGO.

Strong evidence of their absence is given by the immense volume and type of traffic over Moore and Brown's command nets during the battle. Traffic was incessant, chaotic, and almost entirely devoted to retaining control of the situation. Circumstances forced Moore to use his bandwidth in an attempt to save his battalion—which he did—but in so doing he was prevented from using his C2 system to command: to develop and communicate a new concept that would allow him to impose his will on his enemy.

This case is an excellent example of a good C2 system that could not cope with a less-than-well-thought-out command concept. Given the manner in which the C2 system was employed, it failed Moore. It was not organized to peer into the jungle and tell him much about the nature of the enemy, or of its plans, intentions, and likely response to Moore's actions. And, unlike van Creveld's ideal commander, Moore had not "discovered what it [the C2 system] could not do and then proceed to do it nonetheless."

This case provides an excellent opportunity to discuss an extreme example of how our theory deals with a concept that is proven badly wrong by events. Whereas Moore promulgated (notionally) a concept predicated on a lengthy search for and envelopment of a weak and fleeing enemy force, he was quickly presented with evidence that the enemy was neither weak nor fleeing. Although our assessment focused on the failure of Moore's initial concept, we should note that, under extreme time pressure, Moore developed, articulated, and executed at least three separate command concepts over a 36-hour period while in the defense: (1) in shifting forces to keep the LZ open during the insertion of his battalion, (2) in organizing for the attack during the next morning (he expected an attack and had his soldiers execute a "Mad Minute" at dawn, which effectively pre-empted and spoiled the main NVA attack), and (3) in subsequently deciding to preserve and withdraw his force to An Khe.

In our terms of reference, having failed to achieve his initial objectives, Moore deftly crafted and successfully executed three command concepts that (1) were transmitted to and understood by all, (2) were within the capabilities of his C2 system to support, (3) segued from one to another based on his (correct) anticipation of enemy intentions, and (4) were triggered to follow as they did by events that Moore anticipated.

Similarly flawed assumptions underlie the failure of the battle in the next case study, a battle that was structured against available intelligence, rather than in line with it.

# Chapter Eight

# STRUCTURALLY DEFICIENT: MONTGOMERY AT MARKET-GARDEN

Go, Sir, gallop, and don't forget that the world was made in six days. You can ask me for anything you like, except time.

—Napoleon Bonaparte
at Borodino, 1812

## BACKGROUND

In early September 1944, some 200 days ahead of the schedule originally constructed by the D-Day planners, two Allied army groups in France were in a race to the German border. The chief impediment to further advances now stemmed not from enemy action but from the difficulty of resupplying such an enormous effort.

The question in front of the Allied High Command was how to exploit this success. The choices available to General Dwight Eisenhower were rather limited because his strategic reserve consisted of an airborne army rather than of conventional forces.[1] Finding a way to exploit this force had become a priority among the planning staffs. In mid-August, a combined Allied Airborne Headquarters was created and began to plan for a major airborne operation designed to vault Allied forces over the "West Wall" (German fortifications on the western border of Germany) and into Germany.

By the time the first Allied ground patrols neared the German border, the Allied Airborne Headquarters had created and discarded 18

---

[1]The units were the U.S. XVIII Airborne Corps, consisting of the 82d and 101st Airborne Divisions, the British 1st Airborne Division, and the Polish Parachute Brigade.

separate plans. Five had reached the stage of detailed planning. Three had progressed almost to the point of launching. But none had matured. In most cases, fast-moving ground troops were about to overrun the objectives before an airborne force could be put into the fight.[2]

Most of these plans concentrated on getting some part of the Allied armies across the Rhine, which was the most significant natural barrier between them and the German heartland. A plan (code-named COMET) to use one and one-half airborne divisions to seize the Rhine crossing at Arnhem, in the Netherlands, was rejected by Eisenhower in early September as not having enough forces to do the job. The day it was canceled, Field Marshal Bernard Montgomery approached General Eisenhower with a much more ambitious attempt to force a Rhine crossing at Arnhem:

> The new plan was labeled Operation MARKET. Three and a half airborne divisions were to drop in the vicinity of Grave, Nijmegan, and Arnhem to seize bridges across several canals and the Maas, Waal [Rhine] and Neder Rijn [Lower Rhine] rivers. They were to open a corridor more than fifty miles long leading from Eindhoven northward. . . . In a companion piece named Operation GARDEN, ground troops of the Second British Army were to push from the Dutch-Belgian border to the Zuider Zee, a total distance of ninety-nine miles. The main effort of the ground attack was to be made by XXX Corps from a bridgehead across the Meuse-Ascaut canal a few miles south of Eindhoven on the Dutch-Belgian frontier.[3]

MARKET-GARDEN had two main objectives: first, to get across the Rhine, and second, to capture or neutralize Germany's industrial heartland, the Ruhr Valley. The strategic rationale behind MARKET-GARDEN centered on providing an opening for large ground formations to get into and maneuver on the North German Plain (see Figure 8.1).[4]

---

[2]Charles B. MacDonald, *The United States Army in World War II: The Siegfried Line Campaign*, Washington, D.C.: OCMH, 1963, p. 119.

[3]MacDonald, p. 120.

[4]See James Huston, *Out of the Blue: US Airborne Operations in World War II*, West Lafayette, Indiana: Purdue University Press, 1972, Chap. 1.

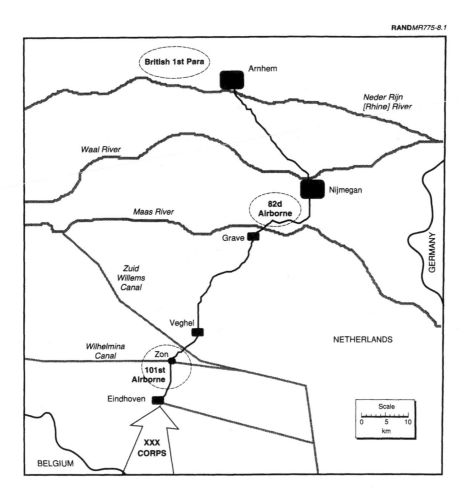

**Figure 8.1—MARKET-GARDEN Operational Concept**

For a variety of reasons, Allied intelligence rated the capability of German forces to oppose this operation as low.  German units throughout the western sector were reeling back toward Germany in various stages of disarray, and planners estimated their ability to provide organized resistance as negligible.[5]  The several warnings that this might not be the case were dismissed as "not credible":

> Despite these warnings, the general view appeared to be as recounted after the operation by the British Airborne Corps.  This was that "once the crust of resistance in the front line had been broken, the German Army would be unable to concentrate any other troops in sufficient strength to stop the breakthrough."  Although the British XXX Corps would have to advance ninety-nine miles, leading units "might reach the Zuider Zee between 2–5 days after crossing the Belgian-Dutch frontier."[6]

During the early days of September, the German Army High Command (*Oberkommando des Heeres*, or OKH) had directed the First Parachute Army, commanded by General Kurt Student, to deploy its cadres to Holland to organize troops fleeing from the debacle in France.  OKH also directed the Fifteenth Panzer Army, commanded by Major General Gustav von Zangen, to withdraw deep into Holland from the Scheldt estuary near Antwerp to reorganize, refit, and stiffen the defense of the northern corridor into Germany.  Both of these armies were subordinated to Army Group B, commanded by Field Marshal Walter Model, whose headquarters, in Osterbeek, lay less than 400 meters from a planned British drop zone.[7]

Chance was to work in the Germans' favor.  In the general confusion surrounding the headlong withdrawal from the Seine, the Germans' ability to monitor and direct the movement of their forces was severely degraded.  One result of this confusion was that, in early September, German forces began piling up in Holland.  In addition to the division-plus of the First Parachute Army and the two divisions of the Fifteenth Panzer Army that made it across the Scheldt, Model, on September 3, ordered the Fifth Panzer Army, which was retreating from the confusion in France, to detach the 9th and 10th SS Panzer Divisions to move to the vicinity of Arnhem for refit and reorganiza-

---

[5]MacDonald, 1963, p. 121.

[6]Headquarters, British Airborne Corps, *Allied Airborne Operations in Holland*, cited in MacDonald, 1963, pp. 122–123.

[7]MacDonald, 1963, p. 126.

tion.[8] In all, the Germans had approximately three panzer division-equivalents scattered throughout the corridor, with two additional panzer divisions in and around Arnhem.

## THE PLANS

Montgomery lobbied hard for the resources with which to execute his plan. He persistently importuned Eisenhower for the authority to supersede all other offensive operations with preparations for MARKET-GARDEN. Finally, Eisenhower promised to deliver a thousand tons of supplies per day from September 10 through October 1. Montgomery promptly set the date for the attack as September 17, giving General Lewis H. Brereton's staff but seven days to plan and prepare for this operation. Brereton's staff immediately began working out the myriad details necessary to move three divisions and all their gear 600 miles into a combat zone.

Brereton appointed Commander British 1st Airborne Corps, Lieutenant-General F.A.M. "Boy" Browning, as commander of the airborne force until such time as it had conducted a link-up with the British Second Army. After that, the airborne force would come under the command of General Sir Miles Dempsey, the 2nd Army Commander:[9]

> On 10 September, immediately after the cancellation of Operation COMET, the commanders of the three airborne divisions were summoned to Montgomery's headquarters to be briefed by Lieutenant-General Browning, who was both Deputy-Commander of the 1st Allied Airborne Army and Commander, British Airborne Corps. . . . When given his orders by Montgomery, Browning was told that the Second Army would be up to Arnhem in two days. Feeling some reservation about this optimism, he made the reply famous in Cornelius Ryan's best-seller, "I think we can hold the bridge for four days but I think we may be going *a bridge too far*."[10]

---

[8]MacDonald, 1963, pp. 135–136.

[9]W.F.K. Thompson, "Operation Market-Garden," in Philip de Ste. Croix, ed., *Airborne Operations*, London: Salamander, 1979, p. 110.

[10]Thompson, 1979, pp. 110–111 (emphasis added).

In planning for the operation, Brereton's staff had to perform four important analyses.

The first was whether to conduct a daylight or a nighttime drop. At night, the flak would be easier to avoid, but the Luftwaffe, swept from the skies during the day by Allied aircraft, had an excellent night-fighting capability. For both ease of navigation and assembly, as well as for air cover, the decision was made to drop during daylight hours.[11]

The second problem was that of the flight routes and formation into the objective area. Long serial formations expose trailing aircraft to alerted enemy gunners, whereas flying the entire formation en masse places so many planes in the same piece of sky that it is difficult for enemy gunners to miss. Brereton's planners decided on a compromise. They developed two routes into the target area—a northern and a southern route. Half the serials would fly by each route.

The third problem was which drop zones (DZs) to select. Doctrine dictated that the best DZ was the objective itself. The problem with the Arnhem objective was that the target (a bridge in the middle of a large city) was in a built-up area. Because glider-borne troops were essential to the success of the plan, Brereton's planners selected a drop zone for the Arnhem operation that was six to eight miles from the bridge over the Neder Rijn but that could handle glider landings.

The fourth problem was how to deliver the force—one lift per day or two. This was the subject of intense debate between the leadership of Troop Carrier Command and the ground commanders. It highlighted the fact that the real problem with making a success of MARKET-GARDEN was not that there were too many bridges but that there were too few aircraft.[12] The ground commanders wanted two lifts per day, to build up combat power on the ground as fast as possible. Troop Carrier Command maintained that two lifts per day placed too much stress on man and machine, and would result in the decimation of their forces. Brereton sided with Troop Carrier Command, despite the fact that planners projected that, with one lift

---

[11]MacDonald, 1963, p. 130.

[12]Thompson, 1979, p. 112.

per day, it would take more than four days, even in perfect weather, to deliver all of the forces into the objective area.[13]

In traversing the 60-mile corridor from Eindhoven to Arnhem (the GARDEN operation), XXX Corps would have to cross three major rivers: the Maas, Waal, and Neder Rijn, plus three major canal systems (see Figure 8.1). The airborne forces (the MARKET operation) would have to seize and hold crossings over all these obstacles. The 101st Airborne was to seize bridges at Zon and Eindhoven in the south; the 82d Airborne Division was to capture bridges over the Maas at Grave, the Waal at Nijmegan, and the Maas-Waal Canal; and the British 1st Airborne was to capture the bridge over the Neder Rijn at Arnhem and establish a bridgehead north of the river sufficient to pass XXX Corps through to the Zuider Zee.[14]

The drop plan was as follows:

- On D-Day, main combat elements of each division—the three regiments of the 101st and 82d, and the three infantry brigades of the British 1st Airborne—would be dropped in.

- On D plus 1, the remainder of the British 1st Airborne was to reach Arnhem, the 101st was to get its glider regiment, and the 82nd was to get its Division Artillery.

- On D plus 2, the Polish Parachute Brigade was to be dropped outside Arnhem, the 82d would get its glider regiment, and the 101st would get its Division Artillery.

- On D plus 3, the remainder of all divisions were to arrive.[15] In all, about 34,000 troops were to be dropped by parachute, and 13,781 were to be landed by glider.

While airborne planning was under way, planning and preparation for GARDEN proceeded apace. General Brian Horrocks, commander of XXX Corps, had to solve the problem of moving 20,000 vehicles along a single highway within a 60-hour period. He planned to do this behind an intense curtain of aerial and artillery bombardment,

---

[13]MacDonald, 1963, p. 132.

[14]MacDonald, 1963, p. 131.

[15]MacDonald, 1963, pp. 132–133.

hoping to achieve a quick breakthrough and penetrate the German defenses below Eindhoven.

As for the Germans, they clearly knew that something was up. German sources reveal, however, that Model's headquarters had not a clue as to the true nature of the Allied plan.[16]

## THE BATTLE

On September 17, between 1230 and 1400 hours, nearly 16,000 troops were dropped and airlanded into the objective areas. The first day's drops went extremely well. By day's end, the 101st Airborne at Eindhoven had captured all of their bridges intact, except for the bridge over the Wilhelmina Canal at Zon, which was blown up as leading elements of the 101st approached it. The 82d at Grave captured all of their D-Day objectives, including bridges over the Maas at Grave and over the Maas-Waal Canal at Heuven, and was moving to secure the high ground to the northwest at Grosboek, which dominated the Nijmegan area.

Meanwhile, the British 1st Airborne jumped into a hornet's nest at Arnhem: into the midst of two half-strength panzer divisions. Having been completely surprised, the Germans reacted commendably to the airborne operation. After fleeing Osterbeek one step ahead of the deploying British paratroopers, Field Marshal Model ordered General Wilhelm Bittrich to move the 9th Panzer Division to protect the Arnhem and Nijmegan bridges. The Reconnaissance Battalion of the 9th Panzer Division stopped three of four British units attempting to reach the Arnhem bridge; Lieutenant-Colonel J. D. Frost's 2nd Parachute Battalion was the only unit to reach its objective. In turn, after heading south some distance toward Nijmegan, the Reconnaissance Battalion was nearly wiped out by Frost's unit as it attempted to recross the bridge back into Arnhem.

The distance of the British DZ from the objective necessitated a dependence on tactical radio communication for command and control at Arnhem. The near-complete failure of the British 1st

---

[16]MacDonald, 1963, p. 136.

Airborne's tactical radio nets forced their commander, General R. E. Urquhart, to command his division by racing about in his jeep in an attempt to maintain contact between his scattered units. During one dash across no-man's land, he encountered a series of German patrols and was trapped for nearly 18 hours in an attic in Arnhem, unable to move.[17]

By the end of September 17, as the airborne units were consolidating and digging in, awaiting the second day's drops, General Horrocks' XXX Corps had fought its way north to within six miles of the 101st at Eindhoven. Fighting along the single road was fierce, and although Horrocks' lead units made significant advances, they did not succeed in penetrating the Germans' lines. Kicking off again in the morning of the 18th, Horrocks' lead units made contact with the 101st at the Wilhelmina Canal near Zon late that evening. British Engineers immediately began bridging operations; by morning, they had erected a tank-capable bridge over the canal. The road was now clear to Nijmegan.

As the lead units of XXX Corps roared up the highway toward the 82d's position at Grave, the situation of Urquhart's 1st Airborne Division was becoming more and more precarious. Field Marshal Model had managed to get enough combat power into the fight at Arnhem to bottle up the British. The 2nd Parachute Battalion was still holding the north end of the Arnhem Bridge, but the rest of the division had withdrawn 4 miles to the east, into a defensive perimeter around the British DZ. The absence of a tactical communications link out of the perimeter made coordination of close air support ineffective. Additionally, there was no way of warning Troop Carrier Command that the preplanned drop zones for resupply were now in German hands. After the second day, the bulk of the daily resupply was dropped into German positions.

Meanwhile, bad weather was causing serious delays in the buildup of combat power in the objective area. The Polish Parachute Brigade, earmarked for Arnhem, arrived two days late, as did Major General James M. Gavin's glider-borne infantry regiment, the 325th. Several attempts by the 82d to take the Nijmegan Bridge on the 19th and

---

[17]Thompson, 1979, p. 120.

20th were stymied by well-sited German defenses guarding the southern approaches to the bridge—helped significantly by the fact that Gavin's division was still at half-strength.

On the afternoon of Wednesday, September 20, Gavin's 504th Parachute Infantry Regiment assault-boated across the mile-wide Maas River in leaky canvas boats and managed, under heavy fire, to capture the northern end of the bridge and to disable the demolition charges the Germans had set. Simultaneously, the Grenadier Guards assaulted and captured the south end of the bridge. The road to Arnhem was finally open.

As success grew in the south, the situation of Urquhart's division at Arnhem worsened. By Thursday, September 21, Frost's battalion at the bridge had been almost annihilated, and the toehold on the bridge had been lost. At the same time, radio contact was finally established between the British perimeter near Osterbeek and the lead elements of XXX Corps. For the first time, General Dempsey, the 2nd Army Commander, was able to discover the predicament of the British 1st Airborne.

Friday, September 22, was a difficult day. Weather again prevented reinforcement and resupply from the air, and XXX Corps made only slight progress toward Arnhem from Nijmegan, although reconnaissance units made it to the Neder Rijn, opposite Osterbeek. The British 1st Airborne was forced into an ever-tighter perimeter, and German resistance to XXX Corps' advance was strengthening.

On Saturday, facing mounting casualties and very difficult resistance, General Dempsey convinced Montgomery to order the 1st Airborne to abandon its positions and withdraw south across the Neder Rijn. Of the 10,095 "paras" who jumped into Arnhem, only 2,163 made it back across the river (Figure 8.2).

## COMMAND AND CONTROL

The Allied forces conducted extensive rehearsals and briefings of Montgomery's plan. Montgomery's intent was well understood by all key leaders. The chief defect in the command and control arrangements was that the overall ground commander, General Dempsey, had no system by which he could receive an indication

RAND*MR775-8.2*

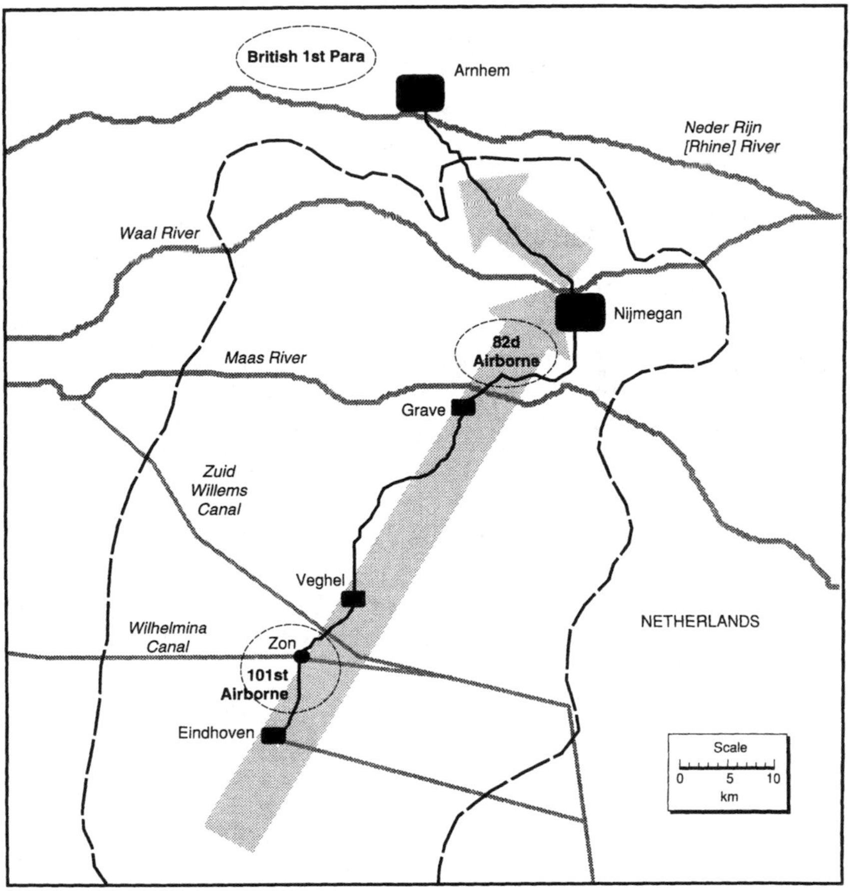

**Figure 8.2—XXX Corps' Progress, September 17–23, 1944**

that the plan was flawed. The British radios at Arnhem were the sole means of communication with the forces advancing from the south. They worked so poorly that the division commander could not communicate even with his *own* forces and was almost captured as a result, which took him out of the battle for over 18 hours.

The entire operational and strategic concept depended on getting across the Rhine and into Germany, an objective that was probably unachievable after the end of the third day. There were several indicators that could have informed Montgomery that his concept was unworkable or unraveling, but because the operational plan did not provide for identifying and communicating these indicators, the Allies continued to pour men and materiel into the objective area, unaware that the prize was no longer for the taking.

## MONTGOMERY'S COMMAND CONCEPT

Embedded in the MARKET-GARDEN planning were many assumptions, most of which concerned especially the ability—or rather the inability—of the Germans to form an organized resistance to the assault. Assuming, however, that weather and the enemy had behaved, an ideal command concept, if explicitly written, might look as follows:

### I. ABOUT THE ENEMY AND HIS PLANS:

1. The enemy [the Germans] currently has no more than 4,000 troops in the Eindhoven-Arnhem corridor. He does not suspect our intentions.

2. The enemy is expected to resist XXX Corps' breakthrough attempt fiercely. However, once the corridor is penetrated, resistance along it will collapse.

3. You [U.S. and Allied troops] should expect the Germans to attempt to reinforce and assist the troops garrisoning the bridge areas at Eindhoven, Nijmegan, and Arnhem with at most 3 infantry battalions and 50 to 100 armored vehicles.

### II. ABOUT OUR FORCE DISPOSITIONS AND PLANS:

1. Our strategic objective is to invade Germany and force a peace by the end of 1944. Our tactical objectives are to (a) cut off the remaining German forces in Holland and force their surrender, (b) outflank the West Wall, and (c) establish a salient across the Rhine to enable a drive into the North German Plain.

2. Over a 3-day period, we shall drop and airland 3-1/2 airborne divisions in three separate areas to seize and hold bridges over major water barriers along the Eindhoven-Arnhem corridor.

3. We shall isolate the Eindhoven-Arnhem corridor area with air and artillery bombardment to facilitate the advance of XXX Corps. XXX Corps will link up with the 101st Airborne at Eindhoven by the end of D-Day, the 82d Airborne will be at Nijmegan by the end of D-Day plus 1, and the British 1st Airborne will be at Arnhem no later than D-Day plus 4.

4. Once across the Neder Rijn at Arnhem, Second Army will exploit this breakthrough by advancing to the Zuider Zee.

## III.  ABOUT CONTINGENCIES:

1. Given the enemy situation, we should be able to capture and hold the six key bridges. However, if XXX Corps cannot break through, the airborne forces will be stranded and will be defeated in detail. A swift advance by XXX Corps will prevent this outcome. Therefore, priority of air and artillery fires will be to XXX Corps units in contact with German defenders, with the objective of destroying them.

2. The insertion and sustainment of our airborne forces is a paramount consideration. If we can get them in and keep them supplied, we should succeed. Therefore, we will select our drop zones, routes, and timing to reduce risks during delivery, even if that imposes additional burdens (such as distance or the absence of surprise) on the delivered forces.

3. If communications with units break down, we shall proceed with our deliveries and resupply operations according to our original plans.

4. Our domination of the air will be used to prevent the Germans from reinforcing their forces in the corridor and to compensate for any unforeseen resistance.

## ASSESSMENT

Given the assumptions embedded in the plan, MARKET-GARDEN might be considered a reasonable gamble. It offered a very attractive prospect: the opportunity of ending the war in 1944. Nonetheless, the judgment that we must render is that, even if the above assumptions were correct, as a command concept the plan was badly flawed: there was no provision for informing Dempsey or moving to a contingency plan if the original plan did not work. The chief problem with the concept was that all of the structural weaknesses of the plan were counterbalanced by the intelligence estimate of the Germans' inability to resist. Given that the estimate turned out to be wrong, verifying this assumption should have engaged every sensor and collection asset in the theater. The plan collapsed from its inherent shortcomings:

> In drawing up its plans, XXX Corps gave little credence to Dutch warnings on how easy it would be for quite small parties of Germans to hold up the advance or interrupt the lines of communication along the single road, much of it on an embankment from which, along considerable stretches, tanks could not deploy. Nor do the warnings of the Dutch Resistance of increasing German strength in the area seem to have been given the weight they deserved.[18]

The Allies gambled that strategic surprise could overcome the extremely limiting terrain, dependence on good weather, poor tactical communications, insufficient airlift assets, and suboptimal location of the drop zone at Arnhem. Neither intelligence nor the C2 system could support the validation of this concept—before or after its implementation began. The true situation at Arnhem—the key to success of the operation—could only be verified by a physical link-up. The result of this conceptual error was the destruction of the British 1st Airborne Division as a fighting force.

---

[18]Thompson, 1979, p. 100.

# Chapter Nine

# SUMMING UP: COMMAND CONCEPTS AND THE HISTORICAL RECORD

If I always appear prepared, it is because before entering on an undertaking, I have meditated for long and have foreseen what may occur. It is not genius which reveals to me suddenly and secretly what I should do in circumstances unexpected by others, it is thought and meditation.

—Napoleon Bonaparte, 1812

Our spectrum of cases and concepts is broad enough to show that viewing C2 from the perspective of command concepts is consistent with the historical record: A commander's ideas and his ability to express them *are* reflected in the demands placed upon the C2 system. The case histories also illuminate the part of C2 that is most poorly handled by the dominant cybernetic paradigm (see Chapter One, "The Foundations of Existing C2 Theory"). When fully articulated, command concepts are generated over a long time scale, deemphasizing the reactive aspect of C2 that is the focus of the cybernetic approach.

## THE RELEVANCE OF THE THEORY

In thinking about the processes that go on among command staffs prior to battle, we have offered a "minimal" hypothesis in which we try to capture as much as possible of this communication over time in a single construct: the commander's concept of battle. Our aim is to shift attention away from C2 support of weak commanders and doctrine-driven operations and toward support of creative, visionary commanders preparing their forces for battle.

Nimitz's concept was so close to the mark and so well understood by his subordinates that the burden on the C2 system (between Nimitz and his subordinates) immediately before and during the battle was minimal. It is interesting to speculate how well the system would have supported his concept if the concept had been flawed. In this regard, we offer a few qualified observations.

First, Nimitz's objective was limited and well-specified, and was based on extremely good intelligence information about enemy dispositions and intentions. It is likely that he would have required an overwhelming quantity of contrary information to be convinced that the Japanese fleet was, in fact, not on its way to Midway. If that were the case, and in view of the vast distances involved, it is unlikely that his intelligence capability would have permitted him to assimilate, decide, and react appropriately. This fact reinforces our proposition that the quality of Nimitz's fundamental idea, which was based on relatively few key pieces of confirmed information, was critical to the effective functioning of his C2 system. Given that his command concept was basically correct, his system was capable of supporting it. If it had been widely off the mark, his C2 system would have failed him—but probably his available forces would have failed him too.

Guderian's concept seems to have been based less on an anticipation of the precise movements of the French army than on an understanding that violent offensive movement into France from an unexpected direction was certain to produce an opportunity to rout the French. Less certain of when that opportunity would occur than Nimitz was, Guderian positioned himself on the battlefield to recognize the opportunity when it arose and to capitalize on it immediately. In this sense, his concept was based on an understanding that a strategic advantage could arise out of the consistent application of superior tactical doctrine, rather than on an expectation of the juxtaposition of a particular correlation of forces.

Guderian's C2 system turned out to be unsupportive of his command style. It could not tell him when his opportunity arose; it did not allow him to verify or refute his concept. Guderian's response was to minimize the load on the system by transmitting only essential information to his Chief of Staff and to rely on the superb tactical ability of his subordinate commanders to operate with minimal guidance and maximum initiative. Guderian compensated for the inability of his C2 system to inform him by being present personally at what he envisioned would be the decisive point on the battlefield. He clearly had given thought to "discovering what it [C2 system] could not do and then proceeded to do it nonetheless." Had Guderian simply relied on the capabilities of his C2 system, the German offensive would likely have stalled at the Meuse. Additionally, had Guderian been badly wrong in his concept, there is little that his C2 system could have done to rectify the matter.

Schwarzkopf's military problem in DESERT STORM was to develop a command concept that would offer a relatively short war, limited casualties, and a decisive defeat of Iraq's military and military potential. Schwarzkopf's C2 system was significantly more sophisticated than that of either Nimitz or Guderian. He had the unprecedented ability to focus on any part of the battlefield from his command bunker. Yet this ability did little to inform him about the Iraqi forces' ability to resist, their willingness to fight hard, or their ability to respond to or anticipate his planned envelopment.

Ultimately, Schwarzkopf's concept was informed by his own experience and judgment. In this case, however, his C2 system was critical to the successful *execution* of his concept. It was able to inform him of the Iraqi forces' reaction to his offensive, and this information was critical to the timing of the various phases of the attack. Without his C2 system's ability to cue him that the Iraqis were crumbling faster than expected, it is likely that a greater portion of the Iraqi army would have escaped and that his strategic objective would not have been achieved.

MacArthur's command concept at Inchon was executed in the near absence of a C2 system. His fundamental idea was planned, rehearsed, and executed solely on the basis of his understanding of the overextension of the North Korean Army and the strategic impact of a landing deep in their rear area. Unlike Nimitz, whose similarly brilliant concept was executed in his physical absence, MacArthur had a front-row seat on his command ship. He watched stolidly as his operation unfolded according to plan, then laconically invited his commanders to get a cup of coffee when it became apparent that the landings were a success. His entire concept was based on his certainty that he would achieve strategic surprise and that resistance to the landings would be minimal.

If MacArthur's calculations had been widely off the mark, and there had been, for instance, several divisions of first-quality NKA troops opposing the landing, the burden on his C2 system to get that information to him would have been minimal. Personally observing the action, he would have quickly understood and been able to react by bringing additional sea and air power to bear in the landing areas.

Moore's fight at Ia Drang is an excellent example of an enormously capable C2 system being unable to rescue a failed concept. Although Moore's leadership was superior and his well-trained unit was able to survive a very difficult fight, he did not have the capability to *understand* what was happening outside his perimeter and what actions the NVA were taking to defeat him—and, consequently, what actions he could take to achieve his objective of destroying the NVA force that had raided Plei Me. Even if Moore's concept had been *right*, and he had successfully engaged an understrength NVA battalion upon landing at LZ X-RAY, his C2 system could not have informed him of threats to the accomplishment of his task, nor of how to exploit the initial defeat of his opponent.

Finally, the importance of getting the "idea" right is clearly demonstrated at MARKET-GARDEN. Montgomery's information architecture, while primitive, did provide the intelligence that German armored units were moving into the Arnhem area. However, Montgomery rejected this information, developing his concept instead on the assumption that the information was untrue. Once the operation was launched, the ability of the C2 system to support alteration or cancellation of the operation was nearly nonexistent: The seriousness of the British 1st Airborne Division's situation became apparent only after a physical link-up with advancing Allied units.

The historical cases we have presented suggest that the quality of the commander's ideas is a critical factor in the functioning of C2 systems. If the idea outpaces the system's ability to support it, the system will likely function poorly, and the command function will suffer. If the idea is flawed or if the concept does not identify the information critical to validating it, the system will likely become flooded with extraneous information, as units in battle attempt to make sense of what is happening.

These cases also suggest that the massive improvements in command-and-control-system performance over the past several decades have not altered the reality that human beings have to know what to look for in order to maximize the performance of the C2 system. From our investigation, we can hypothesize that a well-articulated command concept, one grounded in a realistic, accurate assessment of friendly and enemy intentions and capabilities, is likely to place a lesser burden on the C2 system during its execution, thereby enhancing the ability of the C2 system to provide the few, critical pieces of information that the commander so desperately needs—those that could refute the validity of his concept and cause him to alter the concept.

Finally the six cases examined here suggest two additional hypotheses:

- Information that leads to the development of a sound command concept is at least as important as the information that shows whether a concept is valid or invalid.

- A massive flood of real-time information during a battle is unlikely to significantly alter the outcome of the battle; i.e., there are steep decreasing marginal returns to information.

Our preliminary analysis indicates that those commanders who had a prior command concept and an intuitive feel for battle were able to exploit their technical C2 systems to support the pursuit of their command concept and thereby significantly determine their success. This separation between the intellectual performance of the commander and the technical performance of the C2 system should help both in the technical design and evaluation of C2 systems and in the training and development of battlefield commanders.

## IMPLICATIONS OF THE THEORY

The task of formulating command concepts needs to be embedded in warfighting doctrine. The mapping of command concepts to combat plans must be a feature of battle preparation at all levels.

Commanders at all levels should be evaluated by the quality of the command concepts they develop and promulgate *before* battle, not just by their abilities to improvise and orchestrate actions in response to the unforeseen *during* battle. History—and not just the six cases examined here—suggests that preparation, not improvisation, and vision, not orchestration, are the qualities that have most often carried the day in battle.

To design C2 systems to support improvisation and orchestration—as many U.S. technical efforts seem to be doing—may be teaching future battle commanders the wrong lessons. C2 systems should be designed to help commanders—all commanders—develop sound command concepts before battle and promulgate those concepts clearly and fully to all concerned. During battle, the overriding function of C2 systems should be to inform commanders where and how their command concepts might be wrong and in need of alteration.

Thus, communications doctrine must reflect that heavy traffic in the C2 systems during battle is a sign of a failed command concept: If a commander at any level is improvising and orchestrating battle actions over the command nets, that should be interpreted as *prima facie* evidence that the commanded forces were not properly prepared for the circumstances they actually encountered in battle. This does not mean that the commander anticipates *every* situation that his forces will encounter. It does mean that his forces will know what to do in every situation in order to be acting most consistently with the ideas embedded in his concept—and this includes providing him with the specific information he needs to show him that his concept might have to be altered.

This idea has significant implications for training. Instead of striving to push as much descriptive information as possible up through a communications system during battle and waiting for a reply, the subordinate strives to understand and implement the commander's concept. The subordinate is primed to recognize information that may affect the validity of the commander's concept (a force in a place where none was expected, for example), which will cue the commander that something important is happening to his concept.

It is an accepted truism that armies fight as they have trained. In training today, descriptive information and reactive processes dominate the action. The tempo of operations is intense, both to mimic the realities of the modern battlefield and to maximize the use of scarce training resources. The action focuses on the battle itself, and not on the key activities that, prior to the battle, are essential for developing an understanding of the enemy and his intentions, the structure of possible outcomes, and the elements of information that hold the key to mission accomplishment. Thus, U.S. commanders and armed forces may be training more for the failure of command concepts (and, by implication, commanders) than for how to develop concepts that will succeed in the battles of the future.

Ours is emphatically not an anti-technology argument. The theory simply suggests two things:

1. Command concepts that turn out to correctly anticipate developments on the battlefield will place less of a burden on the C2 system (enhancing its responsiveness, among other things).

2. If development, articulation, and execution of command concepts are the essential elements of the C2 process, then C2 systems should, at a minimum, be designed to ensure that they support that process.

Technology—lots of it—may be essential. When things go wrong in battle, the commander must very quickly develop and articulate a new concept, and this may require a massive amount of technological support in the form of sensors, bandwidth, and decision support systems, as the case of Moore at Ia Drang demonstrates.

The U.S. Army, in its developmental concept paper for Force XXI, is grappling with these same issues:

> Clearly, information technology, and the management ideas it fosters, will greatly influence military operations in two areas—one evolutionary, the other revolutionary; one we understand, one with which we are just beginning to experiment. Together they represent two phenomena at work in winning what has been described as the information war—a war that has been fought by commanders throughout history.

> First, future information technology will greatly increase the volume, speed, and accuracy of battlefield information available to commanders.

> Second, future technology will require the Army to reassess the time-honored means of battle command—to recognize that in the future, military operations will involve the coexistence of both hierarchical and internetted, non-hierarchical processes. Order will be less physically-imposed than knowledge-imposed. . . . Such shared information, where, in some cases, subordinates have as much knowledge as commanders, changes the dynamics of leader-to-led in ways yet to be fully explored and exploited.[1]

Army force developers have concluded that, explosive technological change notwithstanding, the command function seems to possess some enduring characteristics. Technology may have significantly altered the physical aspect of war, but the cognitive aspect of command is proving resistant to technological enhancement.

## RECONCILING CYBERNETIC THEORY WITH COMMAND CONCEPTS

The approach to understanding command and control described in the report does not necessarily contradict the cybernetic view. A fusion of the two approaches may not be of practical value, but it could be accomplished conceptually by extending cybernetic models to represent longer time scales. The processes carried out by humans at various C2 nodes depend not only on general doctrine but also on their understanding of the commander's concept of battle. The formulation and transmission of the commander's concept could be considered as the content of the C2 system over the extended period of time prior to battle. If the commander's concept of the impending battle—communicated prior to battle, during less-intense phases— proves powerful in guiding the actions of subordinates during the height of battle, then its effect is to reduce the communications requirements during battle.

---

[1]U.S. Army, Training and Doctrine Command, *Force XXI Operations: A Concept for the Evolution of Full-Dimensional Operations for the Strategic Army of the Early Twenty-First Century*, Washington, D.C.: TRADOC Pamphlet 525-5, 1994, pp. 1-5, 3-5.

## DIRECTIONS FOR FUTURE WORK

At least three directions can be taken for the next steps in the development of this theory.

The first direction is to examine the implications of this theory for the real-world problem of development of C2 system design.  In an era of limited resources, what does the theory indicate for how to think about procurement decisions?  How should trade-offs be made between improvements in raw power and enhancements to overall system flexibility?  How should decision support systems be designed so that they are empowering but not constraining?  What examples from recent history are illustrative of the rewards and pitfalls of making the right or wrong system decisions?

The second direction for additional work takes a different approach to validating the theory:  Conduct a series of interviews or discussions with (a) living commanders from all the services, to reflect their experiences onto our theory and to inquire whether their experiences resonate with or undermine our theory, and (b) doctrine writers and force developers who are currently grappling with the issue of how to use technological advances to enhance force effectiveness.

The third direction is to extend and refine the theory to ensure that it can be generalized over all services and their different media for operations, especially in view of the growing emphasis on joint operations and military operations other than war.[2]

---

[2]Allard (1996 rev.) catalogs many of these difficulties in Chapter 6, "Tactical Command and Control of American Armed Forces:  Problems of Modernization," especially pp. 169–188.

# ALTERNATIVE MODELS OF COMMAND AND CONTROL

In Chapter One, we described some of the current dominant theories of command and control: control theory, organization charts, and cognitive science. In this appendix, we provide brief descriptions of other cybernetic-related or cognitive-science-related theories, discuss deviations from these dominant paradigms, and relate the cybernetic paradigm to technology development.

## COMMUNICATIONS CONNECTIVITY MODELS

The fundamental need for communications significantly constrains the options for both command and control, making communications infrastructure a critical feature of a C2 system. However, describing the communications links and nodes of a fighting force does not suffice to explain, understand, or predict success and failures in command and control. Converting a description of a communications infrastructure into a causal model of C2 requires that the functions at the nodes be represented. Since the traffic over communications channels is most easily described by messages, an obvious approach to modeling the functions of C2 nodes is to represent the functions as generating, consuming, and transforming messages. The result of doing so is, once again, a variant of a cybernetic model, which captures only those aspects of command that can be described as message-processing tasks—only a fraction of the command burden. This limitation biases C2 models toward the reactive aspects of command and, once again, defines the C2 process as a function of how the system is wired together.

## EXPLAINING C2 WITH ORGANIZATION THEORY

Whereas cognitive science concerns itself with the thought processes of individuals, organization theory applies similar tools to model how organizations "think" as a unit. Organization theory and cognitive science have been employed to produce models that can be thought to apply to the command function generally. An example of such a model is the Headquarters Effectiveness Assessment Tool (HEAT), shown in Figure A.1. HEAT treats C2 as an information-management system.[1] Its intent was to provide measures of C2 effectiveness in terms of mission accomplishment. Note the similarity between Figures 1.3 and A.1.

## SOVIET/RUSSIAN THEORIES OF C2

Soviet literature on C2 also appears to favor a combination of cybernetic and cognitive approaches.[2] Figures A.2 and A.3 reproduce process diagrams from Druzhinin and Kontorov, and Ivanov et al., respectively. While the configuration of the boxes and arrows varies, these models are clearly drawing on the same fundamental concepts as the models in Figures 1.2 through 1.4 and A.1.

Both Soviet/Russian models seem particularly appropriate to the Soviet style of war—careful development of a plan that the commander intends to follow faithfully, the military problem being to select the right plan. In this sense, both models seem very appropriate to the process of developing and promulgating a command concept, but woefully deficient in assessing and managing its execution.

---

[1]Defense Systems, Inc., *Theater Headquarters Effectiveness: Its Measurement and Relationship to Size, Structure, Functions, and Linkages*, Vol. 1, McLean, Va., 1982; D. Serfaty, M. Athans, and R. Tenney, "Towards a Theory of Headquarters Effectiveness," in *Proceedings of the JDL BRG C3 Symposium*, Monterey, Calif., June 1988.

[2]V. V. Druzhinin and D. S. Kontorov, *Concept, Algorithm, Decision (A Soviet View): Decision Making and Automation*, translated and published under the auspices of the United States Air Force, 1972; D. A. Ivanov, V. P. Savelyev, and P. V. Shemanskiy, *Fundamentals of Tactical Command and Control: A Soviet View*, translated and published under the auspices of the United States Air Force, 1977.

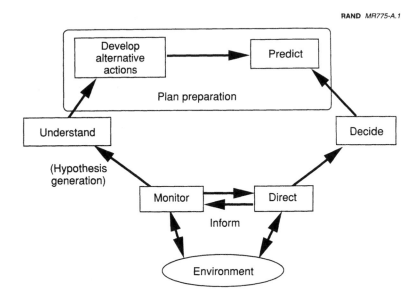

SOURCE: Adapted from D. Serfaty, M. Athans, and R. Tenney, "Towards a Theory of Headquarters Effectiveness," in *Proceedings of the JDL BRG C3 Symposium*, Monterey, Calif., June 1988. Adapted by permission from Daniel Serfaty. Copyright © 1988.

**Figure A.1—Headquarters Process (HEAT)**

## DEVIATIONS FROM THE DOMINANT PARADIGM

The process of displaying C2 models that are essentially cybernetic could be continued at great length (for examples, see Mayk and Rubin, 1988; Stachnick and Abram, 1989; Van Trees, 1989). But the cybernetic approach is not the only option for pursuing a deep understanding of C2. In addition to our theory of C2, which is based on command concepts, there are several other examples of alternative formulations to the cybernetic approach.

Experience in combat and the study of history have left both commanders and scholars with a variety of heuristics for understanding command and control. Beliefs derived using these aids constitute a

RAND *MR775-A.2*

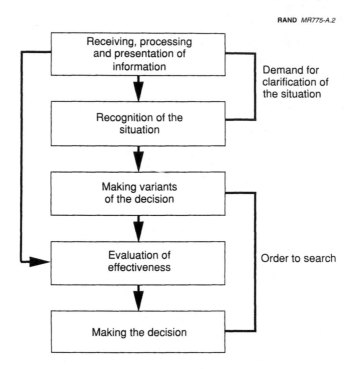

SOURCE: V. V. Druzhinin and D. S. Kontorov, *Concept, Algorithm, Decision (A Soviet View): Decision Making and Automation*, translated and published under the auspices of the United States Air Force, 1972, p. 14.

**Figure A.2—Preparation for Decision**

normative basis for constructing C2 doctrine.  For example, various commanders may believe that effective operations depend on seizing the initiative, maintaining coherent action among all friendly forces, concentrating maximum combat power at the decisive place and time, or "turning inside" the enemy's decision cycle.

To the extent that these doctrinal beliefs can represent the standard by which C2 is judged, they can constitute normative models of the C2 process.  Normative doctrine has explanatory power to the extent that it can be related to, and, in principle, be derived from, a general

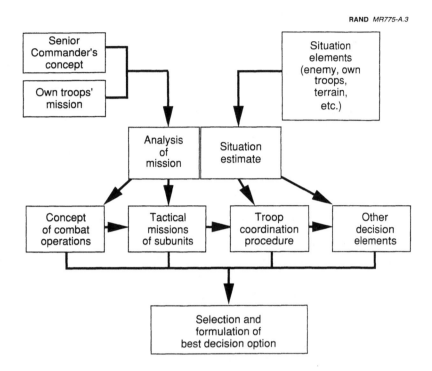

RAND *MR775-A.3*

SOURCE: Adapted from D. A. Ivanov, V. P. Savelyev, and P. V. Shemanskiy, *Fundamentals of Tactical Command and Control: A Soviet View, Soviet Military Thought,* No. 18, translated and published under the auspices of the United States Air Force, 1977, p. 188.

**Figure A.3—The Commander's Decisionmaking Methodology**

theory of combat.    For example, the normative standard of "timeliness" has led to a timeline approach to understanding C2, which has been particularly prominent in strategic command and control (see Figure A.4).    Another interesting alternative is provided by attempts to advance a model of C2 based on the layered protocols of the International Standards Organization (ISO) standard.[3]

---

[3]J. E. Holmes and P. D. Morgan, "On the Specification and Design of Implementable Systems," in Johnson and Levis, 1988, pp. 93–99; Israel Mayk and Izhak Rubin, "Paradigms for Understanding C3, Anyone?" in Johnson and Levis, 1988, pp. 48–61.

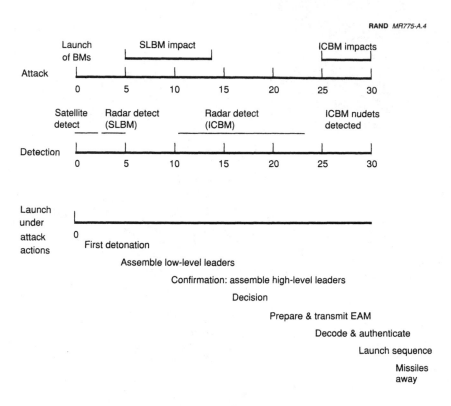

SOURCE: Adapted from H. L. Van Trees, "C3 Systems Research: A Decade of Progress," in Stuart E. Johnson and Alexander H. Levis, eds., *Science of Command and Control: Coping with Complexity*, Washington, D.C.: Armed Forces Communications and Electronics Association International Press, 1989, p. 32. Adapted by permission from Armed Forces Communications and Electronics Association International Press. Copyright © 1989 by AFCEA International Press.

**Figure A.4—Launch Under Attack Timeline**


## TECHNOLOGY PUSH AND THE HEGEMONY OF CYBERNETIC MODELS

The cybernetic approach is clearly the dominant model of C2, judging from both the frequency with which articles about it appear in the command and control literature and the influence it has had on doctrine and technology. This dominance can be explained in part by the ease with which models in this form can be arrived at, starting from a variety of frameworks, and in part by the pragmatics of command and control modeling.

Over the past several decades, a primary use of military modeling has been for hardware acquisition. Cybernetic C2 models are well suited for generating requirements for C2 hardware and infrastructure. The cybernetic approach allows a C2 system to be iteratively decomposed until the arcs represent specific data channels, and the devices of interest are represented by specific boxes in an architectural diagram. This modeling approach is based on and leads to technical characteristics of C2 equipment. In this way, cybernetic approaches to modeling C2 are complementary to technology-driven C2 policy; the need to make and to rationalize C2 acquisitions provides a partial explanation of the success that cybernetic approaches have enjoyed.

Cybernetic models share the common feature that they do not describe processes with long time scales. Cybernetic models represent how C2 systems operate when they are in a responsive mode under significant time pressure. Thus, they may be quite adequate to describe C2 systems for executing standardized procedures in time-critical operations such as maneuver or targeting. In these circumstances, a C2 system operates in a fashion that is very similar to how an automated industrial controller operates.

The process of trying to realistically model command, control, and communications with such models results in a pressure to add complexity by capturing ever more detail in the wiring diagram of the C2 system. This tendency is suggested by Figure A.5, which attempts to integrate indications and warnings (I&W) with command and control activities. Further complexity could be added, for example, by representing C2 processes for all adversaries by cross linkages for jamming, deception, and other forms of information warfare. This process of adding ever-increasing detail leads all such deep modeling in the direction of explanation through simulation, with all the strengths and weaknesses implied by that approach.

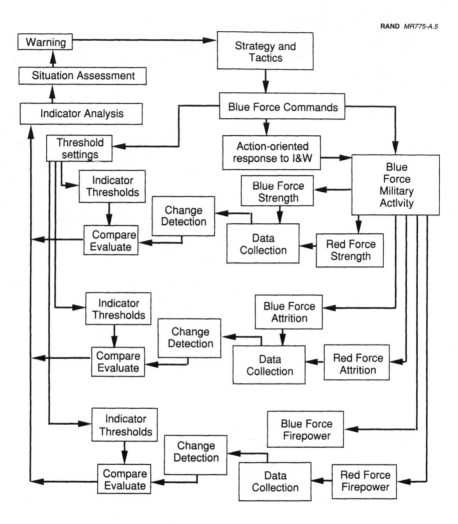

SOURCE: Adapted from A.E.R. Woodcock, "Indication and Warning As an Input to the C3 Process," in Stuart E. Johnson and Alexander Levis, eds., *Science of Command and Control: Coping with Uncertainty*, Washington, D.C.: Armed Forces Communications and Electronics Association International Press, 1988, p. 37. Adapted by permission from Armed Forces Communications and Electronics Association International Press. Copyright © 1988 by AFCEA International Press.

**Figure A.5—Integration of Indications and Warnings (I&W) and Command and Control Activities**

## BEYOND CYBERNETIC MODELS OF COMMAND AND CONTROL

Cybernetic models have legitimate uses, but their exclusive use risks a selective blindness towards those aspects of C2 that are not cybernetic.

Cybernetic models provide a robust basis for understanding *control* functions, but they are inadequate to properly describe *command.* The addition of constructs from cognitive science improves the ability of these models to describe the activity of individuals in the C2 systems. *However, cognitive science is much less adequate to model command decisionmaking than control theory is to model control processes.* Until cognitive science becomes competent to model the full range of the human contribution to these systems, there will be a residual human aspect to command and control that is not captured by cybernetic models.

Human behavior that can be accurately modeled can also be reproduced by an appropriate computer program. *Thus, existing models of C2 would be completely accurate only in the extreme case where the system is completely automated.* Whenever an important human element performs functions that cannot be captured in a computer program, these models will be incomplete. The human aspect of C2 can be modeled when human interaction with the rest of the system can be adequately captured by target lists or explicit plans, or orders. That is, humans behaving in an algorithmic-doctrinal fashion could perhaps be modeled by a cybernetic formalism. But when commanders operate with deeper insight than can be captured by "expert systems"—or when humans from different "boxes" interact without defined structure (meeting in hallways, for example)—these models will be powerless to describe what is going on in more than a superficial, mechanical fashion.

Clearly, these models cannot address those aspects of human behavior that are beyond the reach of existing cognitive science, in particular the creative insights of great commanders. In general, human behavior that is not reactive and is beyond standard operating procedures cannot be accurately captured by these models. Further, because these models represent the decisionmaking elements of C2 systems as atomized information processors, confined to their separate boxes, the cybernetic approaches fail to consider the human cultural element of command and control. Cybernetic and cognitive models represent the interaction of C2 elements, and the decisionmaking that occurs in individual nodes. There is no place in these models to describe processes occurring among groups of officers.

In addition, cybernetic models are poorly suited for representing those human processes that operate over longer time scales. For example, it is known that staffs that have had adequate time to learn to function as a unit have very superior performance to those in which discrete individuals execute doctrinal procedures.

Various time scales are associated with different phases of combat. An armed force that has made adequate preparations will be vastly superior to one that is forced to react automatically at the time of greatest combat intensity, and its demands upon C2 systems will be different. Cybernetic models appear to be specialized to the shortest time scale, the most intense phase of combat, but do not address processes that take place over longer time scales—processes that may put a premium upon gathering information and clearly promulgating plans throughout the force. *They address communication over space, but do not address communication over time.* Similarly, driven by both changing technology and the biases of this modeling paradigm, dominant approaches to C2 emphasize high-intensity quick-reaction aspects of battle command, but provide less structure for understanding longer-time-scale processes of preparation and readiness.

Although we do not pursue the implications of combining quick-reaction aspects and longer-time-scale processes, such a reconciliation could be approached using "playbooks," terse messages that can convey complex meanings. In this framework, the commander's command concept, when fully articulated to subordinates, could be thought of as a complex "play" that assists subordinates in understanding the salience of information obtained during the battle without communicating with higher authority, except to alert the commander of failures in his concept. Transmitted during times when communications bandwidth is not in high demand, the command concept (playbook) would reduce the demands for bandwidth in the heat of battle as long as that concept remained valid.

Albright, John, John Cash, and Allan Sandstrum, *Seven Firefights in Vietnam,* Washington, D.C.: Office of the Chief of Military History (OCMH), 1970.

Allard, C. Kenneth, *Command, Control, and the Common Defense,* New Haven: Yale University Press, 1990.

———, *Command, Control, and the Common Defense,* Washington, D.C.: National Defense University, Institute for National Strategic Studies, 1996 rev.

Appleman, Roy E., *South to the Naktong, North to the Yalu,* Washington, D.C.: Center for Military History, 1961.

Battelle Columbus Laboratories, "Taped Conversation with General Hermann Balck, 12 January, 1979," Columbus, Ohio: Battelle Columbus Laboratories, 1979.

Brown, Rex V., "Normative Models for Capturing Tactical Intelligence Knowledge," in Stuart E. Johnson and Alexander H. Levis, eds., *Science of Command and Control: Coping with Complexity,* Fairfax, Va.: Armed Forces Communications and Electronics Association (AFCEA) International Press, 1989, pp. 68–75.

Buckley, John L., *An Analysis of the United States Army Command and Control Organization in the Pacific Theater: World War II to 1983,* Fort Leavenworth, Kansas: U.S. Army Command and General Staff College (USCAGSC), 1990.

Builder, Carl H., "Is It a Transition or a Revolution?" *FUTURES: The Journal of Forecasting, Planning and Policy,* Vol. 25, No. 2, March 1993, pp. 155–168.

Cash, John A., John Albright, and Allan W. Sandstrum, *Seven Firefights in Vietnam,* New York:  Bantam, 1985 (originally published by the Office of the Chief of Military History, Washington, D.C., 1970).

Chandler, David, *Napoleon's Marshals,* New York:  Macmillan, 1987.

Clausewitz, Carl von, *On War,* translated by Michael Howard and Peter Paret, Princeton:  Princeton University Press, 1989.

Coakley, Thomas P., *C3I:  Issues of Command and Control,* Washington, D.C.: National Defense University, 1991.

———, *Command and Control for War and Peace,* Washington, D.C.: National Defense University, 1992.

Cohen, Eliot, and John Gooch, *Military Misfortunes,* New York:  Free Press, 1990.

Couey, James S., and Randal A. Dragon, *The Impact of Human Factors on Decision Making in Combat,* Monterey, Calif.:  Naval Postgraduate School, 1992.

Currey, Cecil B., *Self-Destruction:  The Disintegration and Decay of the US Army During the Viet Nam Era,* New York:  W. W. Norton, 1981.

Cushman, John H., *Command and Control of Theater Forces: Adequacy,* Washington, D.C.: AFCEA International Press, 1986.

Dearborn, Rebecca D., *An Overview of the Copernicus C4I Architecture,* Monterey, Calif.:  Naval Postgraduate School, 1992.

Defense Systems, Inc., *Theater Headquarters Effectiveness:  Its Measurement and Relationship to Size, Structure, Functions, and Linkages,* Vol. 1, McLean, Va., 1982.

Druzhinin, V. V., and D. S. Kontorov, *Concept, Algorithm, Decision: A Soviet View,* translated and published under the auspices of the United States Air Force, 1972.

Forrestal, E. P., *Admiral Raymond A. Spruance, USN: A Study in Command*, Washington, D.C.: Department of the Navy, 1966.

Freedman, Lawrence, and Efraim Karsh, *The Gulf Conflict, 1990–1991: Diplomacy and War in the New World Order*, Princeton: Princeton University Press, 1993.

Gibbs, M. B., *Napoleon's Military Career*, Chicago: Warner, 1895.

Goodman, I. R., *Evaluation of Combinations of Conditioned Information: A History*, San Diego, Calif.: Naval Ocean Systems Center, 1991.

Gordon, Michael R., and Bernard L. Trainor, *The Generals' War: The Inside Story of the Conflict in the Gulf*, Boston: Little, Brown and Company, 1995.

Greenfield, Kent R., *Command Decisions*, Washington, D.C.: OCMH, 1960.

Guderian, Heinz, *Panzer Leader* (abridged), New York: Ballantine Books, 1967.

Headquarters, Department of the Army, *Decisive Force: The Army in Theater Operations*, Washington, D.C.: FM 100-7, May 31, 1995.

Heinl, Robert D., *Dictionary of Military and Naval Quotations*, Annapolis, Md.: Naval Institute Press, 1966.

——, *Victory at High Tide*, Philadelphia: J. P. Lippincott, 1968.

Helmsley, *Soviet Troop Control*, Oxford: Brassey's, 1982.

Hermes, Walter G., *Truce Tent and Fighting Front [The US Army in Korea]*, Washington, D.C.: OCMH, 1971.

Holmes, J. E., and P. D. Morgan, "On the Specification and Design of Implementable Systems," in Stuart E. Johnson and Alexander Levis, eds., *Science of Command and Control: Coping with Uncertainty*, Washington, D.C.: AFCEA International Press, 1988, pp. 93–99.

Horne, Alistair, *To Lose a Battle: France 1940*, New York: Penguin, 1982.

Huston, James, *Out of the Blue: US Airborne Operations in World II*, West Lafayette, Indiana: Purdue University Press, 1972.

Isely, Jeter, and Phil Crowl, *The US Marines and Amphibious War*, Princeton: Princeton University Press, 1951.

Ivanov, D. A., V. P. Savelyev, and P. V. Shemanskiy, *Fundamentals of Tactical Command and Control: A Soviet View*, translated and published under the auspices of the United States Air Force, 1977.

James, Walter, ed., *Napoleon As a General*, London: The Wolseley Series, 1897.

Johnson, Stuart E., and Alexander Levis, eds., *Science of Command and Control: Coping with Complexity*, Washington, D.C.: AFCEA International Press, 1989.

————, *Science of Command and Control: Coping with Uncertainty*, Washington, D.C.: AFCEA International Press, 1988.

Karnow, Stanley, *Vietnam: A History*, New York: Viking, 1983.

Keegan, John, *The Mask of Command*, New York: Viking, 1987.

————, *The Price of Admiralty: The Evolution of Naval Warfare*, New York: Penguin Books, 1990.

Klein, Gary A., "Naturalistic Models of C Decision Making," in Stuart E. Johnson and Alexander Levis, eds., *Science of Command and Control: Coping with Uncertainty*, Washington, D.C.: AFCEA International Press, 1988, pp. 86–92.

Lawson, J. S., "Command and Control As a Process," *IEEE Control Systems Magazine*, March 1981, pp. 5–12.

Levis, Alexander H., and Michael Athans, "The Quest for a C3 Theory: Dreams and Realities," in Stuart E. Johnson and Alexander Levis, eds., *Science of Command and Control: Coping with Uncertainty*, Washington, D.C.: AFCEA International Press, 1988, pp. 4–9.

Liddell Hart, B. H., *History of the Second World War*, New York: G. P. Putnam's Sons, 1970.

————, *The Other Side of the Hill*, London: Cassell, 1951.

Loendorf, Walter M., *Intelligence Communications: Have We Put into Practice the Lessons Learned in Grenada?* Carlisle Barracks, Penn.: Army War College, 1991.

Lord, Walter, *Incredible Victory*, New York: Harper & Row, 1967.

Ludwig, Emil, *Napoleon*, New York: Boni & Liverwright, 1927.

MacDonald, Charles B., *The United States Army in World War II: The Siegfried Line Campaign*, Washington, D.C.: OCMH, 1963.

Mayk, Israel, and Izhak Rubin, "Paradigms for Understanding C3, Anyone?" in Stuart E. Johnson and Alexander Levis, eds., *Science of Command and Control: Coping with Uncertainty*, Washington, D.C.: AFCEA International Press, 1988, pp. 48–61.

McKee, Alexander, *The Race for the Rhine Bridges: 1940, 1944, 1945*, New York: Stein and Day, 1971.

Moore, Harold G., and Joseph Galloway, *We Were Soldiers Once . . . and Young*, New York: Random House, 1992.

Morison, Samuel Elliot, *History of the United States Naval Operations in World War II*, Vol. 4, *Coral Sea, Midway and Submarine Actions, May 1942–August 1942*, Boston: Little, Brown and Company, 1949.

Naval Communications Command, *Chronological History of US Naval Communications*, Washington, D.C.: Navy Department, 1958.

Naval War College, Dept. of Operations, *Does Copernicus Wear Purple Robes? A Study of New Navy C4I Architecture for the 21st Century*, Newport, R.I.: Naval War College, 1991.

Olmstead, Joseph A., *Battle Staff Integration*, Washington, D.C.: Institute for Defense Analysis, 1992.

Perdu, Didier M., and Alexander H. Levis, "Evaluation of Expert Systems in Decision Making Organizations," in Stuart E. Johnson and Alexander Levis, eds., *Science of Command and Control: Coping with Complexity*, Washington, D.C.: AFCEA International Press, 1989, pp. 76–90.

Potter, E. B., *Nimitz*, Annapolis, Md.: Naval Institute Press, 1976.

Prange, Gordon, *Miracle at Midway*, New York: McGraw-Hill, 1982.

Reynolds, Clark G., *The Fast Carriers: The Forging of an Air Navy*, Annapolis, Md.: Naval Institute Press, 1968.

Robertson, D. C., *Operations Analysis: The Battle for Leyte Gulf*, Newport, R.I.: Naval War College, 1993.

Rolfs, Lawrence L., *Joint Air Combat in the Close Air Battlefield*, Fort Leavenworth, Kansas: USCAGSC, 1990.

Rothbrust, Julian K., *Guderian's XIX Panzer Corps*, London: Cassell, 1974.

Ruoff, Karen L., Roger Thompson, Nicholas R. Todd, and Michael J. Becker, "Situation Assessment Expert Systems for C3I: Models, Methodologies, and Tools," in Stuart E. Johnson and Alexander Levis, eds., *Science of Command and Control: Coping with Uncertainty*, Washington, D.C.: AFCEA International Press, 1988, pp. 118–126.

Sanders, Charles W., Jr., *On Command: An Illustrative Study of Command and Control in the Army of Northern Virginia, 1863*, Newport, R.I.: Naval War College, Dept. of Operations, 1991.

Schnabel, James F., *Policy and Direction: The First Year [The US Army in Korea]*, Washington, D.C.: OCMH, 1971.

Schwarzkopf, H. Norman, and Peter Petre, *It Doesn't Take a Hero*, New York: Bantam, 1992.

Semon, David J., *Combat Leadership—Trouble in the Nineties?* Maxwell AFB, Alabama: Air War College, 1990.

Serfaty, D., M. Athans, and R. Tenney, "Towards a Theory of Headquarters Effectiveness," in *Proceedings of the JDL BRG C3 Symposium*, Monterey, Calif., June 1988.

Sheehan, Michael J., *Selecting a Subset of Stimulus-Response Pairs with Maximal Transmitted Information*, Monterey, Calif.: Naval Postgraduate School, 1992.

Speer, W. H., *Back to Basics: A Five-Dimensional Framework for Developing and Maintaining a High-Performing Battalion or Brigade Staff*, Fort Leavenworth, Kansas: USCAGSC, 1984.

Stachnick, Gregory, and Jeffrey M. Abram, "Army Maneuver Planning: A Procedural Reasoning Approach," in Stuart E. Johnson and Alexander H. Levis, eds., *Science of Command and Control: Coping with Complexity*, Washington, D.C.: AFCEA International Press, 1989, pp. 129–137.

Stamps, T. Dodson, and Vincent J. Esposito, eds., *A Military History of World War II*, Vol. I, West Point: U.S. Military Academy, 1970.

Sun-Tzu, *The Art of Warfare*, trans. Roger T. Ames, New York: Ballantine, 1993.

Tabak, D., and A. H. Levis, "Petri Net Representation of Decision Models," *IEEE Transaction on Systems, Man, and Cybernetics*, Vol. SMC-15, No. 6, 1985.

Thompson, W.F.K., "Operation Market-Garden," in Philip de Ste. Croix, ed., *Airborne Operations*, London: Salamander, 1979.

U.S. Army, Training and Doctrine Command, *Force XXI Operations: A Concept for the Evolution of Full-Dimensional Operations for the Strategic Army of the Early Twenty-First Century*, Washington, D.C.: TRADOC Pamphlet 525-5, 1994.

U.S. Joint Chiefs of Staff, *Department of Defense Dictionary of Military and Associated Terms*, Washington, D.C.: Office of the Joint Chiefs of Staff, JCS Pub. 1, January 1986.

————, *Doctrine for Joint Operations*, Washington, D.C.: Office of the Joint Chiefs of Staff, Joint Pub 3-0, February 1995.

Van Creveld, Martin, *Command in War*, Cambridge, Mass.: Harvard University Press, 1985.

Van Trees, Harry L., "C3 Systems Research: A Decade of Progress," in Stuart E. Johnson and Alexander H. Levis, eds., *Science of Command and Control: Coping with Complexity*, Washington, D.C.: AFCEA International Press, 1989, pp. 24–44.

Warner, Oliver, *Great Sea Battles,* London:  Spring Books, 1970.

Wohl, J. G., "Force Management Decision Requirements for Air Force Tactical Command and Control," *IEEE Transactions on Systems, Man, and Cybernetics,* Vol. SMC-11, No. 9, September 1981, pp. 618–639.

Woodcock, A.E.R., "Indication and Warning As an Input to the C3 Process," in Stuart E. Johnson and Alexander Levis, eds., *Science of Command and Control:  Coping with Uncertainty,* Washington, D.C.:  AFCEA International Press, 1988, pp. 32–47.

Woodward, Bob, *The Commanders,* New York:  Pocket Books, 1992.